Albert Howard Low

TECHNICAL METHODS
OF
ORE ANALYSIS

Elibron Classics
www.elibron.com

Elibron Classics series.

ISBN 0-543-98574-1 (paperback)
ISBN 0-543-98573-3 (hardcover)

This Elibron Classics Replica Edition is an unabridged facsimile
of the edition published in 1919 by John Wiley & Sons, Inc., New York.

not to follow directions exactly. In carrying out a method, the alert operator sees a short-cut and takes it, or a "better way" occurs to him and he introduces it in the place of the one given. There would be no ultimate harm in this (since all methods fall short of perfection) if the operator would only take the time to investigate and determine the real value of his ideas. In some cases he might discover that his supposed improvement was spoiling a good method, and he would come to agree with the author of the method who had himself probably gone over the same ground. I have seen methods of my own thus modified, and ideas hastily adopted whose incorrectness I had previously demonstrated by careful investigation.

I would, therefore, caution operators to accept directions as they stand until they have demonstrated the value of proposed changes to a certainty.

Some of the methods in the following collection have been devised by myself, mainly on the basis of previously well-known facts, some are compilations of the work of others, and some are modifications of existing methods. I have endeavored to give proper credit in all cases.

While I have aimed at correctness, many shortcomings will undoubtedly be found, and I can only hope that, such as it is, my work may prove of some value to those interested. I shall be grateful for any suggestions, or descriptions of improved methods, to be incorporated in a possible future edition.

My thanks are due to Dr. W. F. Hillebrand for valuable information and suggestions.

A. H. Low.

Denver, Colorado, May, 1905.

PREFACE TO THE FIRST EDITION.

THIS book is primarily intended as an aid to the technical chemist, but it is hoped it may also prove useful to the student desiring to become acquainted with technical methods.

It is a common experience with technical chemists to receive a sample of material with instructions to return the percentage of some constituent whose technical determination is more or less unfamiliar under the given conditions. In such a case the chemist has recourse to his books, and too frequently is quite unable to find a method that is exactly adapted to the material in hand, or that begins at the beginning and tells him just how to proceed. He is thus left to work out his own salvation, possibly at the expense of much valuable time.

In this book an attempt has been made to supply the want thus indicated by describing methods that are adapted to the cases most likely to be met in practice, although it is sometimes practically impossible to devise a short technical method that will meet every probable case.

It has been my aim to make the descriptions so minute and complete that if the operator will follow them exactly he can scarcely fail to obtain satisfactory results. But herein lies a difficulty. There seems to be a tendency among technical chemists

PREFACE TO THE EIGHTH EDITION.

It has been found necessary to again revise this book. A number of new methods have developed, notably those for Molybdenum, Potassium, Tungsten and Uranium, which have been placed in the Appendix. New matter has been inserted in the text at various points, and errors as far as found, corrected. Some of the older methods, no longer preferred for regular use, have been allowed to remain, as being possibly useful in special cases or situations.

A. H. Low.

Denver, Colorado, Feb., 1919.

To My Wife,

Whose encouragement has ever lightened my labors,

This Volume is Dedicated

PRESS OF
BRAUNWORTH & CO.
BOOKBINDERS AND PRINTERS
BROOKLYN, N. Y.

TECHNICAL METHODS

OF

ORE ANALYSIS

BY

ALBERT H. LOW, B.S.

Graduate of Mass. Institute of Technology; formerly Chief Assayer, United States Mint, Denver; formerly Chief Chemist, Boston and Colorado Smelting Works, Argo; Member of American Chemical Society; Member of Society of Chemical Industry; Member of Western Association of Technical Chemists and Metallurgists; Member of Firm of Von Shulz and Low, Denver, etc., etc.

EIGHTH EDITION, REVISED AND ENLARGED

TOTAL ISSUE, TEN THOUSAND

NEW YORK

JOHN WILEY & SONS, INC.

LONDON: CHAPMAN & HALL, LIMITED

1919

PREFACE TO THE THIRD EDITION.

TECHNICAL methods of ore analysis are a constant subject of study to the chemists using them. New methods are devised and old ones are frequently modified. In the present revision I have endeavored to make the work as up to date as possible, although old methods are not always omitted, since there might be cases where a chemist, through force of circumstances, would find them desirable.

Slight changes are numerous throughout the book and much new matter has been added. The most notable changes and additions may be found in the chapters on Aluminum, Antimony, Arsenic, Calcium, Lead, Magnesium, Nickel and Cobalt, Tin, Uranium and Vanadium, Zinc, Coal and Coke, and in the Appendix.

I trust those having occasion to use the book will find it satisfactory and reliable.

A. H. LOW.

DENVER, COLORADO, December, 1907.

PREFACE TO THE FIFTH EDITION.

THE present revision has been very thorough and the important changes made are too numerous to refer to specifically. Much of the old matter has been modified, rewritten, or entirely eliminated, and a great deal that is new has been added. The previous style of minute description, although criticized by some, has been adhered to in the main. Brief outlines of methods might suffice for many, but the student or beginner usually requires plenty of detail. Ignorance of some little detail, or lack of attention to it, is a frequent source of trouble. Personally, I have often experienced difficulty with a method that was new to me on account of something being left unsaid in the description. With the opportunity again at hand to condense, I decided not to do so.

I beg to thank those chemists who have kindly furnished me with methods and information. Their names will be found in the text.

A. H. LOW.

DENVER, COLORADO, January, 1911.

PREFACE TO THE SEVENTH EDITION.

In this revision the style of the previous editions has been adhered to. A number of slight corrections and changes have been made throughout the book and there are some additions and omissions, the most notable being in the chapter on Uranium and Vanadium. In the case of uranium the changes were very necessary and important, as this element is now attracting so much attention as a source of radium.

It is hoped that the work will continue to be found worthy of a place among the descriptions of technical methods.

A. H. Low.

Denver, Colorado, June, 1914.

TABLE OF CONTENTS.

CHAPTER I.

CHAPTER II.

CHAPTER III.

CHAPTER IV.

CHAPTER V.

CHAPTER VI.

CHAPTER VII.

CHAPTER VIII.

CHAPTER IX.

CHAPTER X.

CHAPTER XI.

CHAPTER XII.

CHAPTER XIII.

CHAPTER XIV.

CHAPTER XV.

CHAPTER XVI.

CHAPTER XVII.

CHAPTER XVIII.

CHAPTER XIX.

CHAPTER XX.

CHAPTER XXI.

CHAPTER XXII.

CHAPTER XXIII.

CHAPTER XXIV.

CHAPTER XXV.

CHAPTER XXVI.

CHAPTER XXVII.

CHAPTER XXVIII.

CHAPTER XXIX.

CHAPTER XXX.

CHAPTER XXXI.

CHAPTER XXXII.

CHAPTER XXXIII.

CHAPTER XXXIV.

TECHNICAL METHODS OF ORE ANALYSIS.

CHAPTER I.

APPARATUS.

THE standard works on quantitative analysis give descriptions of the usual apparatus required, and a repetition of the information is not contemplated here. I will simply mention a few articles and arrangements that have been found convenient in my laboratory.

1. **Flasks.**—For most of my work I formerly used a 6-oz. flat-bottom flask, ordering a special form having a funnel-shaped mouth, as shown in Fig. 1. The enlarged mouth was for convenience in pouring in weighed material from the scale-pan. More recently I have preferred an 8-oz. flask and have found those manufactured by the Whital Tatum Co. very suitable. Their shape is shown in Fig. 2. Small Erlenmeyer flasks with funnel-shaped mouths, as in Fig. 3, are useful in special cases. An 8-oz. Erlenmeyer, provided with a lip, is especially suitable for the zinc assay. The lip is not for convenience in pouring, but to provide an outlet when the flask is covered with a watch-glass. It prevents the sealing of the top with a continuous ring of liquid, portions of which might be projected and occasion loss.

2. Casseroles.—Flasks are generally prescribed throughout this book, instead of casseroles, for decompositions. They appear to be less liable to mechanical losses of their contents, and for accurate work are generally better adapted and more convenient. In smelter laboratories, where many determinations have to be made daily, casseroles, in connection with a large hot plate, are usually preferred. The usual size is about 3¾ inches in diameter. In many of the cases where I prescribe flasks for

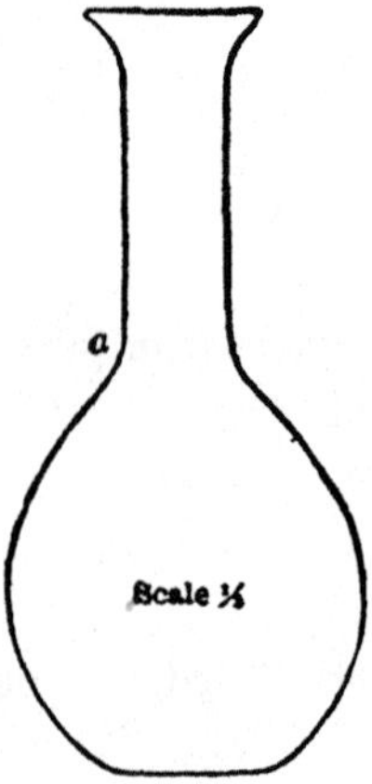

FIG. 1.

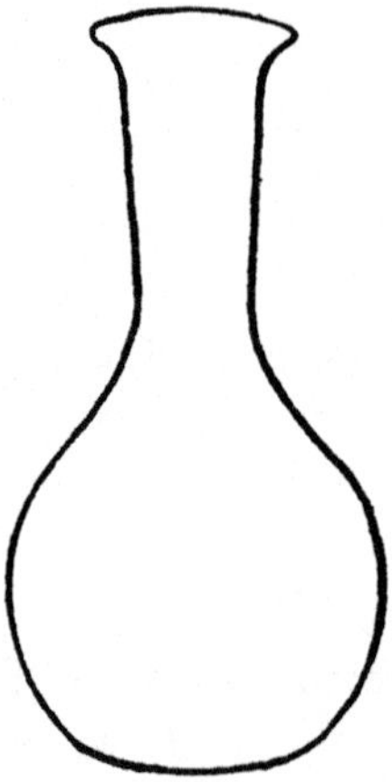

FIG. 2.

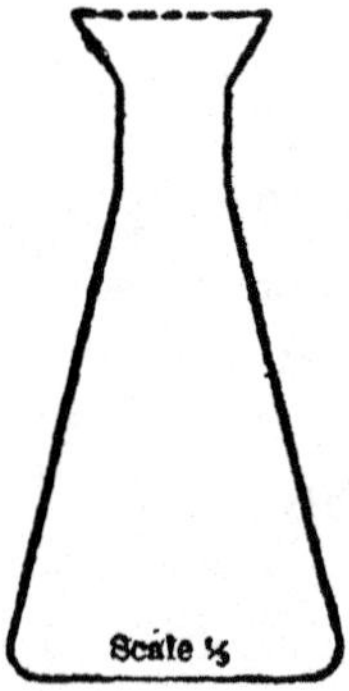

FIG. 3.

decompositions the operator may employ covered casseroles if he prefers.

3. Funnels.—For ordinary routine work I do not use a filter-pump or 60° Bunsen funnels. I take a cheap funnel of about the dimensions shown in the figure and modify it as follows: Cut off the stem a short distance below the neck and melt on a glass tube provided with a loop as shown. In filtering, the loop will cause the liquid to form a continuous column in the tube below and the weight of this column will produce a suction in the filter above. A long stem will give a strong suction, but a length of 10–12 cm. is usually sufficient.* A platinum cone is not used, as the suction is not strong enough to render it necessary. In order to make the filter fit properly, the second fold is made with one side larger than the other, so that the opened filter will have an angle of something more than 60°. The wet filter will then fit the funnel at the top and be too small below. This will prevent entrance of air and increase the filtering surface, so

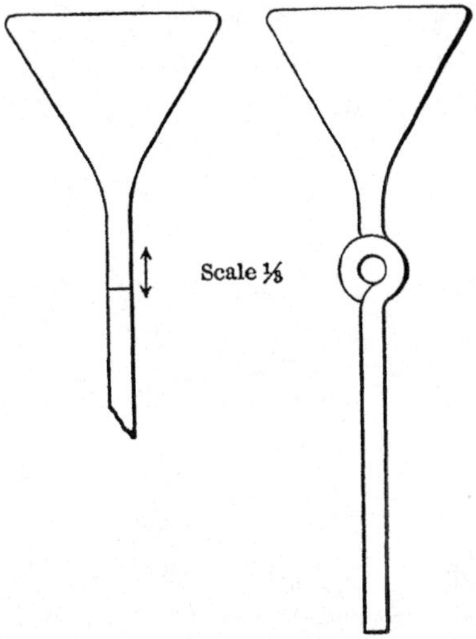

FIG. 4.

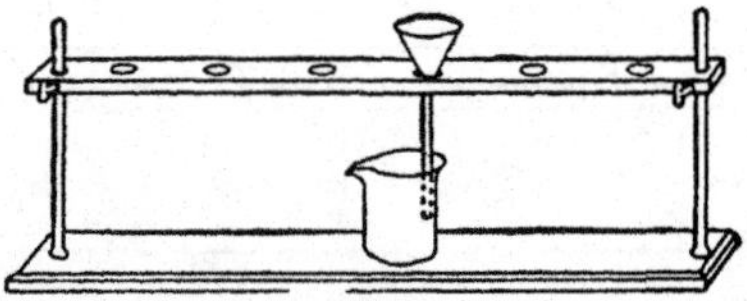

FIG. 5.

that slight suction will effect rapid filtration. After the filter has been placed in the funnel, it is moistened and the top pushed down and fitted with the finger until it is as air-tight as possible. Working by this method, nothing would be gained by using a funnel having an angle of exactly 60°.

4. Funnel-support.—The support shown in the figure is very convenient. It holds six funnels in a compact row, which facilitates manipulation.

* These funnels, and also the other articles mentioned in this chapter, are kept in stock by the Denver Fire Clay Co.

5. Cooling-box.—Considerable time can be saved in copper, lead, and other determinations, where a hot liquid in a flask has to be cooled to ordinary temperature, by employing a cooling box or tank of some sort. The one shown in Fig. 6 consists of a wooden box lined with sheet lead and provided with pipes for the entrance and overflow of cold water. The top board, covering half the opening, is secured in place and has openings for the entrance of 6 flasks. The inner end of each opening is enlarged somewhat and the board is cut away conically underneath, so that a flask which will float will rise into its socket and be in no danger of overturning.

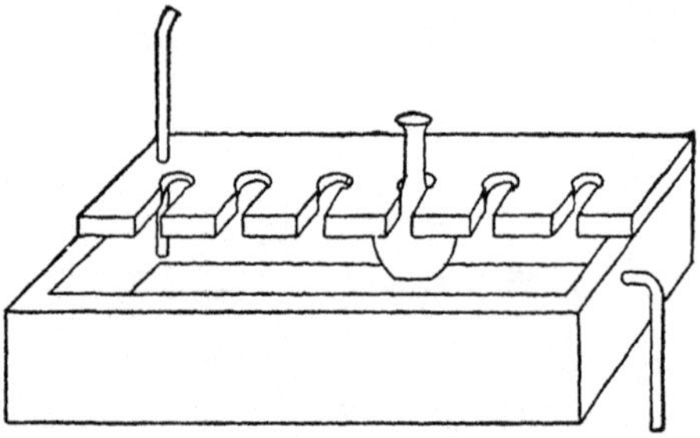

FIG. 6.

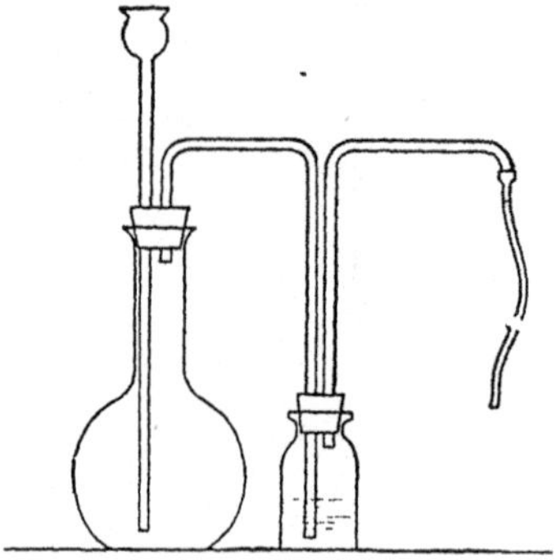

FIG. 7.

6. Hydrogen Sulphide Apparatus.—Notwithstanding all the automatic arrangements that have been devised for generating hydrogen sulphide, I prefer for ordinary use in a small laboratory the simple combination shown in Fig. 7. The flask should be of convenient size, say 16-oz., and the stoppers rubber. The washing-bottle is about half filled with water. It takes only a minute or so to clean and charge this apparatus, and when through using, the excess of gas can always be conducted into water to keep up the supply of hydrogen sulphide water. I have found it convenient, in order to save hood-space, to arrange the generator on the outside and connect the rubber tube with a glass tube passing through the wall.

7. **Measuring-glass.**—For the best work it is indispensable that all reagents should be measured, so that no doubt may exist as to the amount used for any purpose. A 25-cc. measuring-glass is perhaps the most convenient size for constant use. It seems to be an unfortunate fact that almost all measuring-glasses and casseroles are made with the lip on the wrong side for the analytical chemist's use. In the case of the 25-cc. measuring-glass it is a good plan to have a glass-blower make a lip on the other side and also to attach a handle of glass tubing as shown in Fig. 8. The handle keeps the fingers from contact with strong acids, etc., which will inevitably get on the outside of the glass.

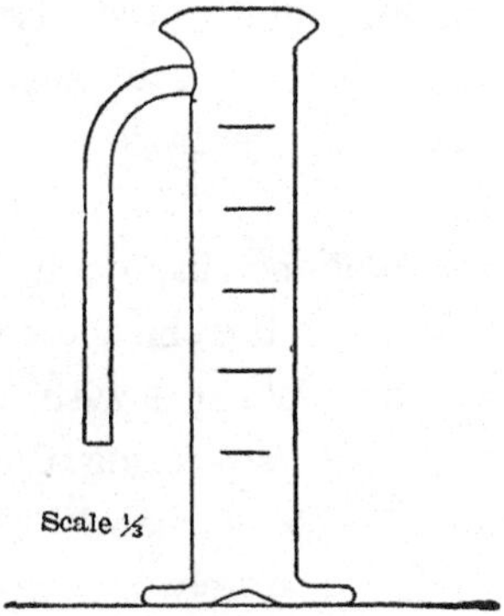

FIG. 8.

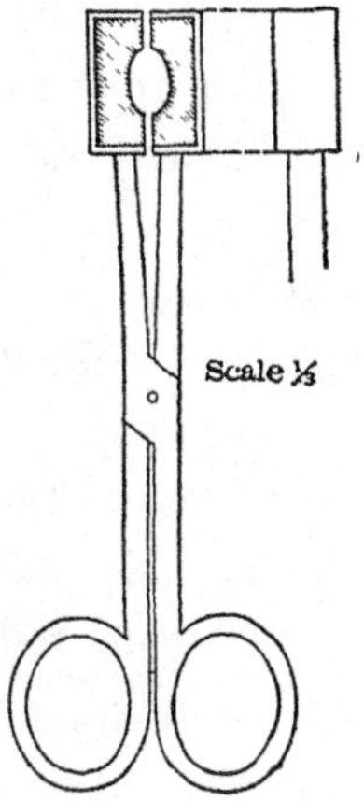

FIG. 9.

8. **Flask-holder.**—The holder shown in Fig. 9 is one of the most useful articles in my laboratory. My first holder came from Holland, and until the Denver Fire Clay Co. copied my model I was unable to obtain a duplicate in this country. It is the best design and the handiest to use of any I have seen.

The portion that grasps the flask is lined with cork cemented in place with sealing-wax.

9. Apparatus for Standard Solutions.—All kinds of arrangements have been devised for facilitating the filling of burettes from the stock bottles of standard solutions. Some of the methods are very convenient and satisfactory, but for the most accurate work with the least apparatus I prefer the following plan:

Keep the standard solutions in large stock bottles of about 9 liters capacity. From the large bottles fill smaller ones, say 1 liter size, as required. Have the burettes enlarged to a funnel shape at the top, as shown in Fig. 10, and keep them stoppered or covered with a cap when not in use. When a burette is to be filled the solution may easily be poured in from the small bottle. By this plan the liquid in a burette can always be mixed before use if desired, and the solution in the small bottle also shaken up. There are no stop-cocks or rubber tubes to keep in order and the solution in the large bottle is disturbed only when necessary to refill the small bottle.

10. Burette Pinch-cock.—For use with solutions that do not attack rubber, I know of no better pinch-cock than the one shown in Fig. 11. It consists simply of a rubber tube plugged with a short section of glass rod or a small bulb. By squeezing the tube around the plug between the fingers slightly, a channel is made for the passage of the liquid. With a properly proportioned tube and plug the arrangement is very satisfactory.

11. Surface of Work-table.—The most satisfactory surface I have tried for the work-bench is asbestos-board, at least $\frac{1}{8}$ of an inch thick,—$\frac{1}{4}$ inch is better. It should be cut to fit the bench and laid in as large sheets as possible. It is usually unnecessary to tack it in place. Instead of covering the entire bench, it may suffice to cover a small section where the operator does most of his work. The outer edge that receives wear should be painted with melted paraffin,—in fact the entire surface may be thus painted, if it is desired to increase the durability. The advantages of the asbestos surface are its comparative indestructibility under the action of corrosive chemicals or heated articles, its softness, thus saving glass utensils, and, if not painted with paraffin, its rapid absorption of spilled liquids. When much soiled or worn it is easily renewed.

Linoleum likewise makes a very durable and satisfactory surface.

12. Rapid Filtration of Gelatinous Precipitates.—Dittrich * has described a method for the rapid filtration of gelatinous precipitates, such as ferric or aluminum hydroxide, which consists in mixing paper pulp with them before filtration. The paper pulp is prepared by violently shaking a piece of filter-paper with a little water in a small stoppered flask or bottle. The paper pulp does not ordinarily interfere with the subsequent ignition of the precipitate.

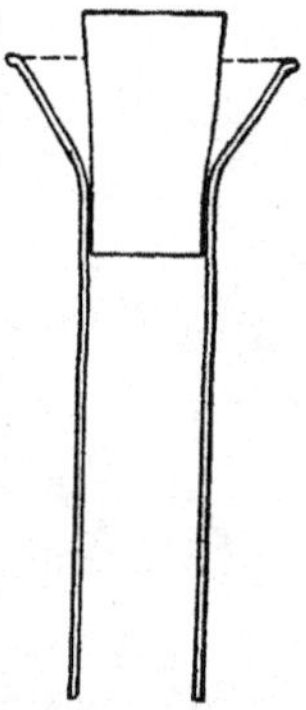

FIG. 10.

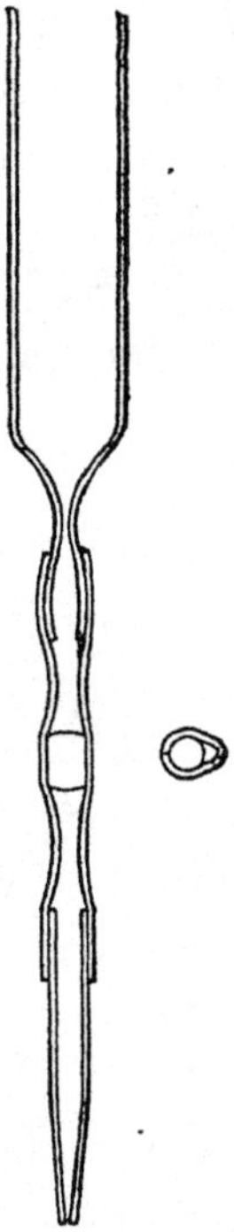

FIG. 11.

I have found the scheme very useful in certain cases. Where it is desired to keep the volume of the filtrate as small as possible, the mixed pulp and water may be first poured into the filter and the drained-off water thrown away. The filtration of the gelatinous mixture stirs up the pulp, which thus appears to be as effective as usual.

* Ber., 37, 1840.

13. Sand-bath. — A "sand-bath," so-called, without the sand, supported on an iron tripod and heated with a Bunsen burner, is an excellent arrangement for flasks and casseroles. The bath is stamped from thin sheet-iron and is of circular form, about 5 inches in diameter and 1 inch deep. A piece of thin asbestos board placed in the bath under a flask, with a properly adjusted flame, will produce a mild diffused heat for gentle boiling to dryness, with the least liability to "bumping," or cracking the utensil after dryness is reached. For a silica evaporation the evaporating-dish may be supported on a scorifier placed in the sand-bath and a fairly strong flame used. This will effect a rapid evaporation without boiling, and the heat is sufficient for the subsequent dehydration of the silica.

14. Boiling Rods.—For the prevention of "bumping" in boiling, glass rods with a short capillary tube on the lower end have been recommended. They have the objection that it is difficult to wash out the capillary opening. To obviate this I have had rods made with a cup-shaped cavity in the end, as in Fig. 12, which permits of easy washing. The larger cavity appears to serve just as well as the capillary tube.

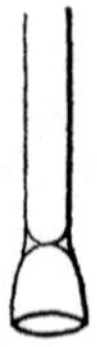

FIG. 12.

15. Wash-bottle for Ammonia, etc.—To prevent the back-flow of corrosive vapors into the mouth I have found the arrangement shown in Fig. 13 very satisfactory. The mouth-tube is of as large diameter as convenient, suitably contracted at the outer end, and extends to the bottom of the flask. The lower end is drawn out to a small opening and is slightly bent away from the exit tube, so as to prevent the blown-in air from entering the latter. It is not bent so much that it cannot be inserted through the stopper. The latter is pierced with a third hole which is covered by a piece of sheet-rubber. The sheet-rubber is cut to fit the top of the stopper and is held in place by the tubing.

It serves as a valve for the third hole when manipulated by pressing and releasing with the finger. To use the bottle, close the valve and blow, taking care to release the valve before discontinuing the pressure. The small opening in the lower end of the mouth-tube, together with the large diameter of this tube, will effectually prevent any residual pressure from forcing liquid

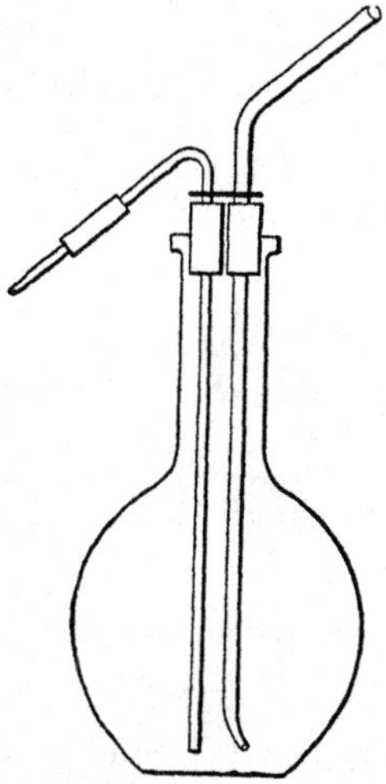

FIG. 13.

back into the mouth. Do not have the small opening *too* small, so as to produce very slow action. The advantage of the above arrangement is its durability, never getting out of order.

16. **Pencil for Glass, etc.**—For marking glass and porcelain the blue pencils of a greasy composition, sold for the purpose, will be found most convenient. They are best sharpened with a coarse file.

CHAPTER II.

ELECTROLYSIS.

The following remarks on electrolysis are intended simply as an aid to those who have to make an occasional electrolytic determination. The apparatus necessary for making a single determination of copper, nickel, or bismuth is described, and a few general directions are given as to manipulation and the attainment of the proper conditions.

Those who desire to pursue the subject further or fit up an elaborate installment are referred to the standard works on electrochemical analysis.

1. **Battery.**—Three ½-gallon Grenet cells, French form. These will give all the electromotive force and current necessary. Both the electromotive force and the current can be sufficiently varied by using from 1 to 3 cells in series. The degree to which the zinc plates are immersed affects the current but slightly. The following solution is used in this battery: For each cell—water, 2 liters; strong sulphuric acid (commercial), 200 cc.; powdered potassium dichromate, 226 grams.

The zinc plates should be amalgamated with mercury. When the battery is not in use, it is best to remove and wash the carbons and zincs and cover the jars with watch-glasses.

2. **Volt-Ammeter.**—Some form of apparatus for measuring the current is quite necessary. Voltmeters and amperemeters

may be purchased as separate or combination instruments. A form of the latter, called the "Student's Volt-Ammeter," manufactured by the L. E. Knott Apparatus Co. of Boston, Mass., will serve very well for ordinary work. It costs about $7.

3. **Electrodes.**—One set of electrodes is sufficient for the determinations mentioned above, as follows: *Cathode.* A plain cylinder of platinum foil, 5 cm. long and 2.5 cm. in diameter. It has a total surface (including both sides) of about 78.5 sq. cm. and weighs about 12.5 grams. A stout platinum wire about 12 cm. long is attached to the top of the cylinder. *Anode.* This is made from a single piece of stout platinum wire. A straight portion about 17 cm. long rises from the center of a circular base, made by coiling the wire closely about itself, so as to form a disc about 2 cm. in diameter. It weighs about 8.5 grams.

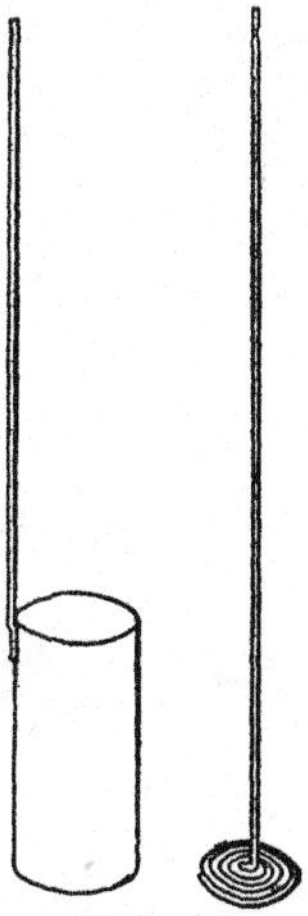

FIG. 14.
Electrodes.

4. **Beaker for Electrolysis.**—This is about 5 cm. in diameter and 9 cm. high. A narrow strip of paper is pasted on the outside at the point indicating a volume of 100 cc. A 4-inch watch-glass split in two serves as a cover, the electrode wires passing through the crack in the center.

5. **Supports for Electrodes and Beaker.**—One of Classen's supports with two clamps (sold by dealers for about $4) will serve very well, or the operator can easily construct a support of wood and a few binding-posts. It is a good plan to have the beaker on a block of wood or other elevated support, so it can easily be raised into place or lowered away from the electrodes as desired. The elevated arrangement of the beaker will also permit of its being supported over a flame if a temperature higher than the ordinary is required.

6. Fig. 15 shows a convenient plan for a single apparatus. The battery and volt-ammeter are placed on a shelf over the work-bench and the electrodes are held by two binding-posts screwed into the edge of the shelf, the connections being made as shown in the diagram. The figure is intended simply to show the arrangement, and in practice the wires may be more or less concealed and connected with convenient switches if desired. The beaker is supported on one or more wooden blocks resting

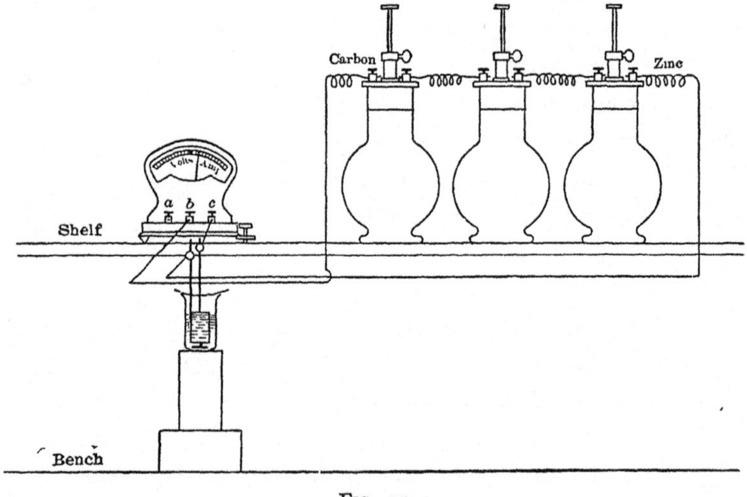

FIG. 15.

on the bench, or a ring-stand may be used. When an electrolysis is finished, the beaker can be held in the hand while the block is removed, and then, with the current still passing, it may be gradually lowered from the electrodes, while the latter are washed with a stream from the wash-bottle. The beaker may then be replaced with one filled with distilled water before the cathode is disconnected. The volt-ammeter indicates amperes to the right and volts to the left of a central zero. The diagram shows the ammeter side in circuit, connections being made with the binding-posts *b* and *c* in series, and the needle inclined to the right. To read volts, connect the battery wires directly with the electrode binding-posts and attach shunt wires from these posts to *a*. and *b*. When the proper connections are made the needle will point to the left.

7. **Conducting an Electrolysis.**—Clean and ignite the electrodes, and, when cool, weigh either one or both as required. Now, by means of the supports and binding-posts, adjust the electrodes and beaker of solution as follows: Place the anode within the cathode with its base perhaps a quarter of an inch below the lower edge of the cylinder. Adjust the beaker and volume of solution so that the anode nearly touches the bottom of the beaker, and the top of the cathode cylinder is about a quarter of an inch above the surface of the liquid. Cover the beaker with the split watch-glass and connect the electrodes with the battery. For copper, nickel, and bismuth the cathode is to be connected with the zinc pole.

Now include the ammeter in the circuit and note the reading. It is not sufficient to know simply the amount of current passing, since the actual effect of this current in the solution is governed by the area of electrode surface over which it is distributed. It is necessary to have a certain amount of current passing into a unit area of cathode surface in order to effect a proper deposition of metal; in other words, the current must have a certain *density*. 100 sq. cm. has been chosen as the unit area to which to refer the current as read in amperes. Thus if a current of 2 amperes is received on 200 sq. cm. of cathode, the current density per 100 sq. cm. is only 1 ampere. Conversely, if 1 ampere of current is passing through 50 sq. cm. of cathode surface the density per 100 sq. cm. is 2 amperes. The symbol ND_{100} is used to express the density of the current per 100 sq. cm. of electrode surface exposed to its action. Knowing the area of the cathode (including both sides) actually immersed in the solution, the current density may be easily calculated from the reading of the ammeter. Thus if the ammeter reads 2 amperes and the cathode has 75 sq. cm. of surface immersed, then $ND_{100} = \frac{2}{0.75}$, or 2.66 amperes.

If one cell fails to give a sufficient current, use two or three in series, as necessary. Under given conditions of resistance a certain electrode tension is required in order to obtain a stated strength of current. This tension may be measured with the voltmeter. The instrument is not, like the ammeter, included in the main circuit, but is placed in a shunt between the electrodes or points in the circuit whose difference of potential is to be measured.

Having attained the proper current conditions, it is best to leave the ammeter in the circuit during the entire electrolysis, so that any variation in the current may be noted and corrected if necessary.

When the appropriate tests show the metal to be all deposited, the beaker may be removed and the electrodes washed as described above, or according to the directions given for the metal being determined.

It is usually best to wash the electrode to be weighed, first with distilled water and then with strong alcohol. It may then be drained a moment on filter-paper, dried by holding over a hot plate, or better, in a drying-oven at about 100° C., and finally cooled and weighed.

A cathode cylinder made of platinum wire gauze has been found to offer some advantages over one made of foil. A freer circulation of the solution is afforded, the time of deposition shortened, and the deposited metal more firmly adherent.

By the use of special apparatus arranged to rapidly rotate

one of the electrodes a very strong current may be employed without injuring the quality of the deposit, and the time required for a satisfactory electrolytic determination thus shortened to 20 minutes or less. Descriptions of such methods may be found in recent electrochemical literature.*

---

* For a short comprehensive article with references, see W. H. Easton in The Chemical Engineer, Vol. 1, p. 386. See also Benner's method, Chap. XIII, 12.

CHAPTER III.

LOGARITHMS.

1. The ordinary calculations of technical analysis may be greatly facilitated by the use of logarithms; furthermore, the liability to error is considerably lessened on account of the fewer figures employed, and for the same reason a calculation is more easily checked over and any error detected.

When logarithms are not used it is a great advantage to have the standard solutions for volumetric work of certain exact strengths, so that their factors will be whole numbers. It is frequently difficult to prepare such solutions, and they are liable also to require troublesome adjustment from time to time. By the use of logarithms the necessity for these hard-and-fast standards is obviated, the calculation with any factor being very short.

Tables of logarithms and antilogarithms are given in this book, but for daily use it is best to have tables mounted on opposite sides of a piece of stout cardboard. They can be purchased in this form of the Franklin Laboratory Supply Co., 15 Harcourt Street, Boston, Mass.

For the benefit of those who are unfamiliar with logarithms, or out of practice, the following remarks are appended:

To *multiply* two numbers, *add* their logarithms. The sum is the logarithm of the product.

To *divide* one number by another. *subtract* from its logarithm the logarithm of the divisor. The remainder is the logarithm of the quotient.

The logarithm of a number is composed of two parts—a positive or negative integral number called the *characteristic* and a positive decimal fraction called the *mantissa.*

All numbers composed of the same figures placed in the same order have the same mantissa irrespective of the position of the decimal point.

The characteristic of a logarithm indicates the position of the decimal point in the corresponding number. When the number has one significant figure to the left of the decimal point, the characteristic of its log. is 0. If there are two figures it is 1, with three figures it is 2, etc. A decimal fraction has a negative characteristic which is equal to the number of places the first significant figure is removed from units; thus, the characteristic of the log. of 0.0469 is -2; of the log. of 0.0046, -3, etc. A negative characteristic is written with the minus sign immediately over it. It may be made positive by adding 10 to it, the fact being indicated by writing 10 with a minus sign after the mantissa.

The above rules may be illustrated as follows:

Number.	Logarithm.
4691	3.6713
469.1	2.6713
46.91	1.6713
4.691	0.6713
.4691	$\bar{1}.6713$ or $9.6713-10$
.04691	$\bar{2}.6713$ or $8.6713-10$
.004691	$\bar{3}.6713$ or $7.6713-10$

The -10 after the mantissa is frequently omitted where the value of the characteristic is obvious.

In many simple operations the position of the decimal point may be seen at a glance without the use of characteristics. For instance, when working with a standard solution where 1 cc. = about 1 per cent. of the constituent sought, then, if 45 cc. have been used and the figure corresponding to the log. obtained (no characteristics having been used) is 4563, it is evident that the percentage is 45.63, and not 4.56 or 0.45.

Four-place logarithm tables are intended to be used with numbers of not more than four figures. Such tables, however, will suffice for most of the calculations of ordinary analytical work.

To find in the table the mantissa of the log. of a number proceed as follows: If the number consists of only 1 or 2 figures, find the number in the extreme left-hand column and the mantissa will be given immediately opposite in the next column to the right headed 0. Thus the mantissa of the log. of 1 or 10 is .0000, of the log. of 12, .0792, etc. If the third and fourth figures of a number are 0, the mantissa will also be in the 0 column as before. If the third figure is an integral number, the mantissa will be found in the same horizontal line, in the column headed with that figure; thus the mantissa of the log. of 124 or 1240 is .0934. If the fourth figure of a number is an integral number, recourse must be had to the column headed "Proportional Parts." Having found the mantissa for the first three figures, there must be added to it the amount found in the same horizontal line in the column headed with the fourth figure in the proportional parts. Thus the mantissa of the log. of 1245 is .0934+.0017, or .0951. Having found a required mantissa, prefix the characteristic according to the rules given above.

The table of antilogarithms gives the numbers corresponding to logarithms, and a number is found from its logarithm in precisely the same manner as a logarithm is found in the table of logarithms. Thus the number corresponding to log. 9.0974 is 0.1251, the characteristic 9 signifying the same as $\bar{1}$ and indicating the position of the decimal point. In the same way log. 1.0974 gives 12.51, etc.

CHAPTER IV.

ALUMINUM.

1. THE technical determination of aluminum (usually required as Al_2O_3) in ores and metallurgical products is a somewhat troublesome proposition. There seems to be no short and satisfactory method that is applicable to complex substances. The usual interfering elements are iron, manganese, arsenic, antimony and phosphorus. Chromium and titanium would prove sources of trouble, but are so rarely present in western lead-smelting ores that they may ordinarily be neglected. There are two general methods of procedure, the direct and the indirect. In the direct method the aluminum is weighed as $AlPO_4$ or Al_2O_3. In the indirect method the aluminum, iron, and perhaps phosphorus are weighed together as Al_2O_3, Fe_2O_3 and P_2O_5, the latter, of course, combined to form a phosphate with the others. The alumina is then found by difference after determining and deducting the weight of the other constituents of the mixture.

2. **Direct Method.**—Treat 0.5 gram of the ore in a platinum dish with 2–3 cc. of strong sulphuric acid and about 20 cc. of strong pure hydrofluoric acid. Evaporate over a water-bath or other gentle heat as far as possible and then cautiously raise the heat until the sulphuric acid is fuming copiously. Allow to cool, add water and a little hydrochloric acid and warm the

mixture until the dish is free from adhering insoluble matter. Now transfer, using as little wash-water as possible, to an 8-oz. flask. Add 6 grams of potassium sulphate, 5 cc. of strong sulphuric acid and one-eighth of a 9-cm. filter. Boil the mixture gently at first, to expel water and hydrochloric acid, then, as the sulphuric acid begins to fume, the heat may be increased as strongly as the prevention of undue foaming will permit, finally with the flask in a holder, over a free Bunsen flame, until any free sulphur is entirely expelled and the separated carbon is completely oxidized, leaving a clean mass or melt with but little free sulphuric acid. Allow the flask to cool in an inclined position to avoid cracking. The object of the filter paper is to reduce any arsenic or antimony to the *ous* condition, and thus render its subsequent precipitation as sulphide rapid and complete.

3. After cooling, add 150 cc. of water and 5 cc. of strong hydrochloric acid and warm the mixture until everything soluble has dissolved. If a trifling residue remains it may usually be neglected. From a larger residue, decant most of the solution through a filter and heat the residue with dilute hydrochloric acid, repeating these operations once or twice if necessary. Calcium sulphate or lead salts may thus be dissolved or greatly reduced in amount. Pour the solutions through the filter, finally transferring any remaining residue, and wash with hot water. Reserve the filtrate. The final residue may consist of a little barium or calcium sulphate or other unimportant substance. If its character is thus recognized it may be negletced, but if this is uncertain and aluminum is liable to be present it is best to treat it further. Ignite filter and contents in a platinum dish until the carbon is consumed. The residue may now be either again evaporated with a little hydrofluoric and sulphuric acid nearly to dryness, or, if silica is apparently absent, fused with a little mixed sodium and potassium carbonates. In either case

the solution of the product, acidified with hydrochloric acid, may be added directly to the reserved filtrate, without regarding any residue of barium sulphate.

4. The liquid is now an acid solution of the various bases. It should not be too acid, and it is therefore safer to nearly neutralize with ammonia and then reacidify with an excess of 5 cc. of hydrochloric acid. Now dilute to about 300 cc. with hot water and pass in hydrogen sulphide to precipitate the metals of that group. Ten minutes will usually suffice for this precipitation, since any arsenic or antimony is in the *ous* condition and will come down quickly. It is well, however, to have the liquid fairly cool at the end, to ensure the complete precipitation of any lead. Filter, washing with water containing hydrogen sulphide. Boil the filtrate until all hydrogen sulphide is expelled and then oxidize the iron to the ferric condition by the cautious addition of a few cubic centimeters of nitric acid to the boiling solution. Now dilute to about 400 cc. with cold water and allow to cool to room temperature. The liquid is now ready for the precipitation of the aluminum as phosphate according to Peters's modification of Wöhler's method,* as follows:

5. Add ammonia until the solution becomes brown or dark red in color, according to the amount of iron present, but still contains no precipitate. Now add 3.3 cc. of hydrochloric acid of 1.2 sp. gr. and 2 grams of sodium phosphate dissolved in water and filtered if necessary. Stir until any precipitate formed is dissolved and the solution becomes perfectly clear again.† Now add 10 grams of sodium thiosulphate, dissolved in water

---

* Blair, Chem. Anal. of Iron, 3rd ed., p. 250.

† Blair's directions. Personally, I have frequently found the solution to become more and more turbid. I then cautiously add more hydrochloric acid until the solution clears, note the extra amount used and increase the sodium thiosulphate at the rate of 10 grams for each 3 cc. of acid required.

(and filtered if necessary), and 15 cc. of acetic acid of 1.04 sp. gr. (This strength may be approximated by adding 10 cc. of water to 5 cc. of the 80 per cent. acid.) Heat to boiling, boil 15 minutes,* and filter as rapidly as possible on an ashless filter. Wash thoroughly with hot water. If the amount of the precipitate is small it may at once be dried, ignited in a porcelain crucible, together with the filter, and weighed as $AlPO_4$. Multiply this weight by 0.4185 to obtain the weight of the Al_2O_3.

If the precipitate is large in amount it should be redissolved and reprecipitated. Rinse it into a beaker, add a little hydrochloric acid to dissolve it, and then pour through the filter, to dissolve what was not rinsed off, and wash the filter thoroughly. Dilute the filtrate somewhat, if necessary, add ammonia in slight excess, and then reacidify with acetic acid in slight excess. Heat to boiling, filter, and wash with hot water. Then dry (I have found this apparently unnecessary), ignite, and weigh as above described. It is always well to again boil the filtrate from the first precipitate of aluminum phosphate for some time and filter off the precipitate formed on a separate filter. Ignite this in a separate crucible (not weighed), and if any $AlPO_4$ is found, brush it in with the main portion. A precipitate of sulphur will thus always be obtained and the presence of aluminum phosphate can be determined only by igniting it. It is best to repeat this boiling and filtering until the ignited precipitate leaves no residue.

6. Method of J. M. Bonsal.—This scheme, which employs a modification of Wohler's method, is described as follows (Pardoe's revision of Furman's "Manual of Practical Assaying," p. 218): Weigh out 0.5 gram of the finely powdered ore and mix with three to five times its weight of the mixed carbonates of sodium

---

* Blair's directions. I have found thirty minutes safer.

and potassium in a platinum crucible. Fuse until the melt is clear. Dissolve the fused mass in hot water and hydrochloric acid, digesting if necessary. Evaporate to dryness and remove silica in the usual way (XXIV, 11). Add to the filtrate from the silica 20 grams of ammonium chloride, roughly weighed, and precipitate the iron and aluminum with ammonia. Filter the precipitated hydroxides and wash with hot water until chlorine is removed (shown by adding silver nitrate to portions of the filtrate, a white precipitate of silver chloride being formed). Remove the precipitate, which consists of Al_2O_3, Fe_2O_3, and Cr_2O_3 (if chromium is present), from the filter and dissolve in as small a quantity of dilute hydrochloric acid (one part HCl, sp. gr. 1.20, and one part H_2O) as possible. It is very important at this point to avoid an excess of hydrochloric acid. Now add to the solution about 5 cc. of a saturated solution of hydrogen sodium phosphate until a slight precipitate appears, and then 2 or 3 drops of hydrochloric acid, or sufficient to clear up the solution. Now add 30 grams of ammonium chloride, roughly weighed, and 15 cc. of a saturated solution of sodium thiosulphate and boil for fifteen minutes. Filter on an ashless filter, wash five or six times with boiling water, dry, place in a porcelain crucible and ignite, gently at first, until all carbon has been burned off, as the aluminum phosphate may fuse and render the combustion difficult. Weigh as $AlPO_4$, multiplying by 0.4185, to obtain the weight of the Al_2O_3.

7. Note on the Above.—The decomposition, as described, is not applicable to sulphides or material containing easily reducible metals. In such cases, or where in doubt, I would suggest the following procedure: Treat 0.5 gram of the ore in an 8-oz. flask (or small covered casserole) with whatever acids are necessary for its decomposition. Usually 10 cc. of strong hydrochloric acid, subsequently followed with 5 cc. of strong nitric acid if sulphides are present, will prove sufficient. Finally, add 5 cc. of strong sulphuric acid and boil to fumes. Now, manipulating the flask over a free flame, expel nearly all the remaining sulphuric acid. Cool, add 30 cc. of water, heat to boiling, and allow to stand, hot, until any anhydrous ferric sulphate is entirely dissolved. Then cool to room temperature or cooler and filter through a 9-cm. filter, washing the residue four times with dilute (1:10) sulphuric acid. Receive the filtrate in an evaporating-dish. Add ammonia in slight excess to neutralize the sulphuric acid, then reacidify slightly with hydrochloric acid and place the solution to evaporate to dryness. Wash the residue on the filter with hot dilute hydrochloric acid, containing some ammonium chloride, until any lead sulphate is removed, then ignite filter and residue in a platinum crucible or dish until the carbon is all consumed. Finally, fuse the residue with the mixed carbonates, as described above, and add the acidified solution of the melt to the portion already evaporating, proceeding from this point as described.

I would also suggest washing the precipitated aluminum phosphate at least ten times, to insure the removal of all soluble salts.

8. Indirect Method.—This is useful in technical work only when the precipitate of mixed iron and aluminum hydroxides can be easily obtained free, or nearly so, from phosphorus and arsenic, as in the case of clay and similar material. If phosphorus or arsenic are liable to be present in more than negligible amount their separation or estimation is involved, and the process usually becomes too tedious for a technical method. Where the method is applicable the procedure may be as follows:

Take 0.5 gram of the substance and decompose it, according to its nature, either with acids or by fusion in platinum with a mixture of 3 grams of sodium carbonate and 2 grams of potassium carbonate. In the latter case fuse until clear, disintegrate the cold melt by heating with water and transfer the mixture to an evaporating-dish. Cover the dish, acidify with hydrochloric acid' and evaporate to dryness to remove silica. If acids alone have sufficed for the decomposition, the solution may be similarly evaporated. Sometimes an evaporation to dryness, in platinum, with hydrofluoric acid mixed with hydrochloric or a little sulphuric acid (moistening the material with the latter before adding the hydrofluoric acid) may serve to decompose the substance and remove silica at the same time. After the evaporation take up in hydrochloric acid and dilute. Dilute the filtrate from the silica, or the solution otherwise obtained free from silica, to from 100 to 300 cc., according to the amount of iron or aluminum apparently present, make slightly alkaline with ammonia and heat to boiling. Keep at a boiling temperature for a short time, then remove from the heat and allow to settle somewhat and filter, washing with hot water. If the amount of the precipitate is large it is best to redissolve it in hydrochloric acid and repeat the precipitation. It is always safer to do this in any case. The final washing should be very thorough, 10 to 20 times with hot water, or until a portion of the filtrate shows no test for chlorine. Finally, dry and ignite the mixed hydroxides and weigh as Fe_2O_3 and Al_2O_3. After weighing, fuse in platinum with a little potassium acid sulphate, take up the melt in dilute hydrochloric acid and determine the amount of Fe_2O_3 volumetrically, thus arriving at the weight of the Al_2O_3 by difference. Phosphorus, arsenic, and the heavy metals are presumably absent from the substance tested. If manganese is present in appreciable amount it should be separated from the iron and aluminum by the basic acetate method (see Fresenius or Treadwell) previous to the precipitation by ammonia.

CHAPTER V.

ANTIMONY.

THE technical determination of antimony in ores, etc., is best made by volumetric methods. The most troublesome interfering element is usually arsenic. Where arsenic is known to be absent or negligible the procedure in the following method may be shortened by omitting the operations for its removal.

1. Before beginning treatment the nature of the material should be considered. Most sulphides and mixed ores yield readily to the acid treatment described below, but oxidized ores rich in antimony may fail of complete decomposition. Such material is easily decomposed by the method described in 6.

2. **Method Applicable to Sulphides and Most Mixed Ores and Low-Grade Oxides.*** — Take 0.5 gram of the finely pulverized ore in an 8-oz. flask, add 5 grams of ammonium sulphate, 1 gram of potassium sulphate, 10 cc. of strong sulphuric acid and one-eighth of a 9-cm. filter. The object of the latter is to provide organic matter for the reduction of arsenic and antimony to the *ous* condition, thus facilitating the solution of the antimony and also rendering the subsequent precipitation of both metals as sulphides rapid and complete. Heat the mixture cautiously at first, over a small free flame, then increase the heat gradually, and finally manipulate the flask in a holder over the full flame of a Bunsen burner. Continue until any free sulphur is all expelled, the separated carbon completely oxidized and the free sulphuric acid almost entirely driven off, leaving a clean melt from which

* A. H. Low, Jour. Am. Chem. Soc., XXVIII, 1715, modified.

ammonium sulphate is slowly volatilizing. If not perfectly fluid add sufficient potassium acid sulphate to make it so. The above treatment may appear severe for a flask, but one rarely breaks, although, at the end, the bottom is apparently almost at low redness. Allow the melt to cool with the flask on its side, as otherwise the solidifying cake may crack it, or spread the melt around on the sides while cooling.

3. After cooling, add 25 cc. of water, 25 cc. of strong hydrochloric acid and 2 or 3 grams of tartaric acid. Cover the flask with a small watch-glass and warm the mixture, but do not heat to boiling lest some arsenic be lost by volatilization, in case it is to be determined. Boiling would also weaken the acid strength. Maintain fairly hot, with occasional agitation, until solution is as complete as possible. Anhydrous ferric sulphate will dissolve more quickly if a little ferrous sulphate be added. Now rinse off the watch-glass with a very little water, add 25 cc. more of cold strong hydrochloric acid and then pass in hydrogen sulphide until any arsenic present is completely precipitated. As the arsenic is in the *ous* condition this will take only a short time, perhaps ten minutes. Some copper may be precipitated, but no antimony or tin. Filter through a double 9-cm. filter that has been moistened with 2:1 hydrochloric acid. Use an ordinary suction funnel. A platinum cone will insure the filter from breaking. Rinse off the delivery-tube with 2:1 hydrochloric acid and remove it. Rinse out the flask several times with the same acid mixture, and wash the precipitate with it also at least six times. Receive the filtrate in a large beaker. If arsenic is to be determined, reserve the precipitate, and any particles adhering to the flask and delivery-tube, to be proceeded with according to VI, 2. Continue with the antimony as follows:

4. Dilute the filtrate with double its volume of warm water and saturate it with hydrogen sulphide. Simple dilution will

cause a precipitation of antimony sulphide, but it is not safe to omit passing the gas also. Filter off the precipitate, which may consist of antimony sulphide mixed with other sulphides, and wash it from six to ten times with hydrogen sulphide water so as to remove nearly all the chlorides. Use an 11-cm. filter. With a jet of hot water, using as little as possible, wash most of the sulphide from the filter into a beaker. Add a little colorless sodium sulphide solution * and warm to effect the solution of the antimony sulphide. Pour through the last filter and wash filter and residue well with hot water containing a little sodium sulphide. One extraction will usually suffice. Receive the filtrate in an 8-oz. flask. Small amounts of antimony sulphide may be dissolved directly on the filter. To the filtrate add 2 grams of potassium sulphate and about 8 cc. of strong sulphuric acid. (The addition of organic matter for reduction is here unnecessary.) Boil as previously described (2) to expel, first the water, then all the free sulphur, and finally most of the free acid. Small amounts of chlorine appear to be harmlessly expelled before the antimony sulphide is decomposed. Cool with flask inclined. Add to the residue 50 cc. of hot water and 10 cc. of strong hydrochloric acid. Heat to effect solution and then boil for a short time to expel any possible sulphur dioxide. Finally, add 10 cc. more of strong hydrochloric acid, cool under the tap, dilute to about 140 cc. with cold water and titrate to the usual pink tinge with a standard solution of potassium permanganate. The solution ordinarily used for iron titrations (2.83 grams per liter) will answer. The iron value of the permanganate

* Dissolve 100 grams of sodium hydroxide in sufficient water and dilute the solution to 1 liter. Saturate 600 cc. with hydrogen sulphide and then add the remaining 400 cc. 5 or 6 cc. usually suffice for the above extraction. Ammonium sulphide may be used, although it dissolves more copper sulphide. A *small* amount of chlorides in the reagents will not cause an appreciable subsequent loss of antimony by volatilization.

multiplied by 1.075, or, the oxalic acid ($C_2O_4H_2 \cdot 2H_2O$) value multiplied by 0.9532, will give the antimony value. A small amount of copper (frequently present) has no influence on the titration. If the solution becomes turbid before or during the titration, add a few grams of tartaric acid to clarify it.

5. **Notes.**—I have been of the opinion that the mixture of sulphates prescribed for the decomposition was better than potassium or sodium sulphate alone, but recent tests have indicated that a little antimony (or arsenic) may volatilize with the ammonium salt. I would therefore think it safer not to use the ammonium sulphate, but, instead, take 3 or 4 grams of either sodium or potassium sulphate. By holding the flask inclined during the decomposition, so that the melt is on the side, instead of the bottom, any bulging of the latter is prevented.

Sometimes an appreciable amount of undecomposed material remains after the above acid treatment of the substance. If this is likely to contain antimony it may be further treated as follows: Proceed as described until after the first filtration, thus obtaining the undecomposed residue and any arsenic sulphide on the filter. Reserve the filtrate and continue with the material on the filter as described for arsenic (VI, 2), until the latter is removed, omitting this step, of course, if no arsenic is present. Ignite the thoroughly washed filter and residue at a low temperature in the thin iron crucible described below (6) until the carbon is all burned off, and then fuse with a little sodium hydroxide. Dissolve the melt in hot water, transfer to a beaker, acidify with hydrochloric acid, and add the solution to the reserved filtrate. Continue with this as described.

Tin is without influence on the antimony result (since it exists in the final solution as stannic sulphate) but may prove a great annoyance, if much is present, owing to its troublesome sulphide. In such a case I have found Clark's oxalic acid method for preventing the precipitation of tin sulphide very satisfactory. Simply dissolve 10 grams of oxalic acid crystals in the filtrate containing the antimony, previous to diluting and passing in hydrogen sulphide. The tin will be kept in solution and occasion no further trouble.

6. Decomposition of Oxidized Material.—In a thin-spun iron crucible of about 60 cc. capacity melt about 8 grams of sodium hydroxide. (I take about three inches of the stick hydroxide, broken into short pieces.) Heat until the moisture is expelled and quiet fusion attained and then allow to cool. It is a good plan to add a very small pinch of potassium nitrate to the melt just previous to cooling to guard against any subsequent reduction of metal from the ore. Weigh 0.5 gram of the ore, transfer it to the prepared crucible, cover with a loosely-fitting porcelain cover and fuse the mixture. Heat very cautiously at first until violent bubbling has ceased, and then with the full flame of a Bunsen burner until quiet fusion is attained. Remove the cover and pour the melt into a clean metallic dish floating in a beaker of water. I use a 2½-inch nickel dish. It is best to cover the hot cake with a small porcelain crucible cover dropping within the dish, to prevent mechancial loss in case the cake cracks and flies apart violently. Place a little cold water in a 5½-inch casserole and set the hot crucible therein; then turn the latter on its side so as to admit the water and heat to boiling. Move the crucible about with a glass rod, and when the outside is clean wash it off so as to remove the crucible with the fingers. The inside of the crucible may still contain some of the undissolved melt. Pour in a little water, acidify with hydrochloric acid, and bring the dilute acid, by means of a glass rod, in contact with any adhering melt. When all is dissolved, wash the solution into the casserole. Now add the detached cake and the crucible cover with the top (provided with a loop) up. Cover the casserole and add about 30 cc. of strong hydrochloric acid, which should prove an ample excess. Heat to boiling and the cake will quickly dissolve. Now remove and wash off the watch-glass, and by means of a bent wire lift out the crucible cover and wash it. As a rule the decomposition is perfect and everything goes into solution in the hydrochloric acid, even the silicic acid. If scales of oxide from the crucible remain, allow to settle, decant the solution and warm the residue with a little strong hydrochloric acid, adding also, if necessary, a few crystals of potassium chlorate. Add the solution to the main portion. Barium might separate as sulphate, but this is easily distinguishable from undecomposed ore. Now add 2 or 3 grams of tartaric acid, and when it has dissolved dilute sufficiently and filter if necessary, which is rarely the case. If gelatinous silica clogs the filter, a few drops of hydrofluoric acid added to the filtering liquid will usually clear it.

7. Treatment of Solution.—Having obtained a hydrochloric acid solution containing all the antimony, together with sufficient tartaric acid, dilute to about 300 cc. with hot water. If gelatinous silica separates, the addition of 1 or 2 cc. of strong hydrofluoric acid will usually cause it to dissolve. Heat nearly to boiling and pass in hydrogen sulphide gas until the precipitation is complete. Filter off the precipitated sulphides and wash them at least ten times with hydrogen sulphide water to remove the hydrochloric acid. Now with a jet of hot water and using as little as possible, rinse the bulk of the precipitate (through a wide-neck funnel) into an 8-oz. flask. Place the flask under the filter and dissolve from the latter any still adhering antimony sulphide with a little warm ammonium sulphide. In place of these operations the entire precipitate, if its amount is small, may be extracted with ammonium sulphide directly on the filter.

To the contents of the flask, which need not form a solution, add 3 grams of potassium sulphate and 8–10 cc. of strong sulphuric acid. This mixture now corresponds to that of the original ore described in **2**. The addition of a piece of filter-paper is unnecessary. Follow the directions given in **2** from this point.

8. Method for Hard Lead, etc.—In the absence of arsenic treat 0.5 gram of the alloy, which may be in moderately coarse particles, precisely as described for ores (2), until the usual melt is obtained. When cold, add 50 cc. of hot water and 10 cc. of strong hydrochloric acid. Heat until solution is complete and then boil off any sulphur dioxide possibly present. Add 10 cc. more of strong hydrochloric acid, cool completely under the tap, dilute to about 140 cc. with cold water and titrate with permanganate as in **4**. Tin, lead and small amounts of copper and iron do not interfere. If arsenic is present it may be removed by volatilization, as follows: Take up the melt in 25 cc. of water and 20 cc. of strong hydrochloric acid. Boil the mixture until only 15 cc. remain. Now add 15 cc. of strong hydrochloric acid, and cool, dilute and titrate as before. A concentration to 20 cc. may fail to remove all the arsenic if much is present, and a concentration to 10 cc. will occasion an appreciable loss of antimony. The procedure is therefore not perfectly satisfactory. The safest way, when arsenic is liable to be present, is to treat the melt as described for ores (**3**). An alloy containing a large amount of copper should be finely divided and treated like an ore (**2**), without modification of the method.

9. Another Method for Hard Lead, etc.—Treat 0.5 gram of the very finely divided alloy in an 8-oz. flask with a mixture of 5 cc. strong nitric acid, 10 cc. of water, and 2–3 grams of tartaric acid. Heat gently until solution is complete and then boil off most of the nitric acid. Add sufficient water (and more tartaric acid if the solution becomes turbid) and pour the mixture slowly into a solution containing 20–25 grams of sodium monosulphide and 10 grams of sodium hydroxide in 300 cc. of water. Warm the mixture, allow the lead sulphide, etc., to settle, and filter, washing with water containing sodium sulphide. Make the filtrate slightly acid with dilute sulphuric acid, allow the precipitated antimony sulphide to settle, and then filter and wash with dilute hydrogen sulphide water. Treat the antimony sulphide from this point as described in 4, or, if arsenic is liable to be present, proceed as follows: Rinse the precipitate from the filter as completely as possible with a jet of hot water into a 400-cc. beaker. Place the beaker and contents under the funnel and dissolve the precipitate still adhering to the filter with a little warm ammonium sulphide solution. Make the mixture in the beaker just acid with hydrochloric acid and then add to it just double its volume of strong hydrochloric acid. Warm the mixture slightly and stir until all the antimony sulphide has dissolved and then pass in hydrogen sulphide to again precipitate any arsenic sulphide that may have also dissolved. Finally, filter through a double filter as described in 3, and continue as described in 3 and 4.

10. **Rowell's Method for Antimony in Ores and Alloys.***— The following method is based on the reaction

$$KBrO_3 + 3SbCl_3 + 6HCl = 3SbCl_5 + KBr + 3H_2O.$$

A method based on this reaction was first suggested by Gyory and modified by Nissensen and Siedler for testing hard leads. The present method is a modification of the latter.

Standard Arsenious Chloride Solution.—0.8236 gram of pure powdered arsenious oxide, dried carefully at 100° C., is weighed

* H. W. Rowell, Jour. Soc. Chem Ind., XXV, 1181.

into a 500 cc. graduated flask. About 5 cc. of 10 per cent. sodium hydroxide solution are added, and the flask shaken and gently warmed until the arsenic is completely dissolved; 5 cc. of strong hydrochloric acid are added and the solution made up to the mark with water. Fifty cubic centimeters of this solution, equivalent to 0.10 gram of antimony, are measured off with a pipette, the delivery of which has been checked against the 500 cc. flask. A solution made in this way and kept in a well-stoppered bottle remains unchanged for a fortnight without oxidation.

The Indicator. — 0.1 gram of pure methyl orange is dissolved in 100 cc. of hot water. Drops of this indicator should be taken out with a tube to avoid the stain left about the neck of the bottle, as this or an old solution is liable to leave a stain and spoil the sharp finish. Methyl blue and other methyl colors stable in hydrochloric acid generally stain the solution, while indigo turns green toward the finish, and the end is not so easily distinguished as with methyl orange. More of a solution of indigo is also required to produce a sufficient color, which makes the test a trifle higher.

The Standard Potassium Bromate Solution is approximately N/20, and is made by dissolving 1.41 grams of the "pure" salt in 1 liter of water. The theoretical quantity is 1.3926 grams, per liter, but the "pure" salt always contains bromide which, however, makes no difference to the test. To standardize the solution, 20 cc. of strong hydrochloric acid are added to 50 cc. of the standard arsenic solution, and the mixture just brought to the boil. Thirty cubic centimeters of the bromate solution are then added, and the titration finished in exactly the same way as in the method below, observing all the precautions. The solution should be standardized every week, as it loses value at the rate of about 2.5 cc. per liter per week.

A one-fifth normal solution may be used for titrating larger quantities of antimony.

Bromine Solution. — Thirty-five cubic centimeters of pure bromine, thoroughly shaken up with 250 cc. of strong hydrochloric acid, makes a saturated solution, and leaves excess of bromine. The required quantity is conveniently measured by dipping in a pipette without any stem below the bulb and so allowing it to fill.

Method of Analysis. — 1 gram of the finely divided ore or alloy, containing not more than 0.15 gram of antimony, is weighed off into a 500 cc. Bohemian beaker. These quantities may be varied, provided that more of the substance can be conveniently dissolved, and that the amount of standard bromate solution required is within the compass of the burette used. To the sample, 25 cc. of concentrated hydrochloric acid and 5 cc. of the saturated solution of bromine in hydrochloric acid are added; the covered beaker is then placed on a warm iron plate, so that the temperature is not high enough to drive off the bromine before complete solution is effected, and occasionally shaken until complete solution takes place.

Antimony oxides, and precipitates of mixed oxides of antimony and tin which do not dissolve readily in this way, may be fused with eight times their weight of caustic soda in a silver crucible at a dull red heat, till the mass turns yellowish-green. The fused mass is dissolved in as little water as possible and transferred to a 500 cc. beaker and the solution acidified with hydrochloric acid and evaporated down to 10 cc., when 20 cc. of hydrochloric acid are added. To reduce the antimony, 3 or 4 grams of fresh sodium sulphite crystals are added, the cover and sides of the beaker lightly rinsed down with water, and the liquid evaporated, with the cover on, to 10 cc., or a little less if possible. Although there seems to be little risk of the antimony oxidizing during the evaporation, the cover is better kept on, as it retards the evaporation very little and often saves a test when it, or one near it, spurts through evaporating too far.

Sodium sulphite is better than sulphurous acid for effecting the reduction, as it raises the boiling point considerably toward the finish and ensures complete volatilization of the arsenic. If more than 2 or 3 per cent. of arsenic are present, 20 cc. of strong hydrochloric acid and 5 cc. of saturated sulphurous acid are added, and the liquid boiled down again.

To the concentrated solution, 20 cc. of strong hydrochloric acid and 40 cc. of hot water are added, the cover and sides of the beaker are rinsed, and the whole is boiled for one minute to remove traces of sulphurous acid. The standard solution of potassium bromate is now run in to within a few cubic centimeters of the necessary amount, with constant and thorough stirring, and at the rate of 30 cc., at most, every 50 seconds. If lead chloride begins to crystallize, the solution must be boiled again, but otherwise two drops of methyl orange solution are added and the bromate run in drop by drop till the color of the indicator is destroyed. The solution should be kept at a minimum temperature of 60° C., and should be thoroughly stirred during the titration so that a local excess of bromate never forms, otherwise some of its value is lost before attacking the antimony. The result is calculated from the equation

$$\frac{(\text{cc. of bromate required by 1 gram sample} - \text{blank})\ 10}{\text{cc. of bromate required by 50 cc. of arsenic solution}} = \%\ \text{Sb}.$$

A blank test should be made occasionally in exactly the same way as above, omitting the sample, and the result (which should not exceed 0.2 cc.) subtracted from each test.

Possible Sources of Error. — The most probable sources of error are the incomplete removal of arsenic or sulphur dioxide.

Oxidation need not be feared if the cover is kept on. The

liquid must not be allowed to evaporate to dryness, as antimonious chloride begins to volatilize at about 195° C., and boils at 220° C.

Lead, zinc, tin, silver, chromium and sulphuric acid have no effect upon the test, but large quantities of calcium and ammonium salts tend to make the result high.

Iron tends to make the results high, but not in as great a ratio as copper. Iron is very slightly reduced by sulphurous acid in a strong hydrochloric acid solution, and if, before adding the sodium sulphite to a test, it is boiled down to as small volume as possible and made up again with cold, strong hydrochloric acid, the effect of iron is almost destroyed. With this precaution, 1 per cent. of iron raises the test about 0.02 per cent., while 5 per cent. has very little more effect.

The Effect of Copper. — Copper is partially reduced by sulphurous acid in a strong hydrochloric acid solution, and, under the conditions of the method given, raises the test in a fairly constant ratio as shown by the following figures:

	Gram Sb.
0.001 gram of copper as cupric chloride in a blank test	= 0.0001
0.005 " " " " "	= 0.0005
0.010 " " " " "	= 0.0012

The quickest and most satisfactory way of obviating the effect of copper is to dissolve the substance in 15 cc. of nitric acid 1 : 2, evaporate just to dryness, and boil for a few minutes with 50 cc. of 1 per cent. nitric acid; allow to settle and pour off the liquid through a fine filter. Add 30 cc. of 5 per cent. solution of ammonium nitrate, boil again, and transfer all the precipitate to the filter, wash two or three times with a hot 5 per cent. solution of ammonium nitrate, and dry the filter and precipitate. Separate the latter from the filter and fuse it and the filter ash together as directed above.

Perhaps as exact a way when the copper amounts to about 1 per cent. is to subtract 0.012 per cent. of antimony for every 0.1 per cent. of copper, the copper being estimated by a separate colorimetric test.

By the above method an estimation of antimony in antimonial lead can be carried out in one hour from the time of weighing.

CHAPTER VI.

ARSENIC.

(See Appendix.)

For the technical determination of arsenic in ores and metallurgical products I have found the following method to be perhaps the most generally applicable and reliable (also see 11):

1. **Method for Ores, etc.***—Treat 0.5 gram of the ore precisely as described for the determination of antimony (v, 2) up to the point, indicated in the text (v, 3), where the arsenic is obtained on the filter as sulphide, the antimony being in the filtrate. Now proceed as follows:

2. Rinse out any sulphide adhering in the flask or on the delivery-tube as completely as possible, pouring through the filter, and then wash the filter and contents well with hydrogen sulphide water. Absolutely complete removal of the hydrochloric acid is unnecessary. Dissolve the arsenic sulphide (this may usually be done directly on the filter), and also any particles of sulphide on the delivery-tube, with warm ammonium sulphide solution and wash with the same solution diluted, receiving the filtrate in the original flask. To the filtrate add 1 gram of potassium sulphate, 3 grams of ammonium sulphate, and 8 cc. of strong sulphuric acid. Boil the mixture to a small bulk and then manipulate the flask over a free flame until all the sulphur is expelled and a clean melt of the mixed sulphates is obtained. Cool with the flask inclined as a precaution against cracking.

* A. H. Low, Jour. Am. Chem. Soc., XXVIII, 1715. Modified.

Any small amount of hydrochloric acid present is expelled during the heating before the arsenic sulphide is decomposed, and therefore causes no loss of arsenic by volatilization as chloride. Take up the cool residue by warming with 50 cc. of water and then boil a short time to expel any possible sulphur dioxide. Now cool to room temperature, add a little starch solution (XIII, 2; this is an acid solution), then, with a little phenolphthalein as indicator, make slightly alkaline with a solution of sodium hydroxide, and again re-acidify slightly with hydrochloric acid. Cool once more if necessary. Finally, add 3 to 4 grams of sodium acid carbonate and titrate with standard iodine solution. Pay no attention to a brownish discoloration toward the end, but proceed slowly until a single drop of the iodine solution produces a strong permanent blue color. If neutral starch liquor is used it need not be added until just before titration.

3. The iodine solution may be prepared by dissolving about 11 grams of iodine in a little water, with the addition of about 20 grams of potassium iodide, and diluting to 1 liter. Standardize with pure powdered arsenious oxide. Dissolve about 0.150 gram by warming gently in an 8-oz. flask with a little sodium hydroxide solution. Dilute somewhat, cool, add starch, neutralize, etc., similarly to the above and finally titrate with the iodine solution. One cubic centimeter of the latter will equal about 0.003 gram of arsenic. Multiply the weight of As_2O_3 taken by 0.7575 to obtain the equivalent weight of As.

A small amount of copper is without influence in the titration provided it remains in solution. If a precipitate of copper arsenite is formed, add a little ammonium chloride to dissolve it.

4. **Method for Lead, Copper, Alloys, etc.**—It will frequently suffice to take 0.5 gram of the finely divided material and treat it as just described for ores. Or, the method described in V, 9, may be followed until the arsenious sulphide, free from antimony,

is obtained on the double filter. From this point follow the directions given in 2.

5. **Pearce's Method, Modified.***—Thoroughly mix 0.5 gram of the finely ground ore in a platinum dish or large porcelain crucible, with 5 grams of a mixture of equal parts of dry sodium carbonate and potassium nitrate. It is a good plan to reserve a portion of the mixed salts for use as a cover. Heat the mass gradually over a Bunsen burner to complete fusion. It is best to use a very low flame at first and take plenty of time so that the mixed salts will melt and soak through the mass before much decomposition occurs; in this way loss of arsenic is prevented with some ores that tend to lose arsenic by volatilization. Finally, heat to the full power of the Bunsen burner until thorough decomposition is effected. Prolonged heating over a blast-lamp is sometimes necessary, especially with oxidized ores containing lead.

6. The melted mass should finally present a smooth and homogeneous appearance if the dish or crucible is taken up in the tongs and given a circular movement. Cool, extract the soluble portion by heating with water until thoroughly disintegrated, and then filter and wash the residue with cold water. Receive the filtrate in a 6-oz. flask. Drop a bit of litmus paper into the flask and then add nitric acid carefully until the solution is plainly acid, simply avoiding a large excess, but if a precipitate has formed always add enough acid to dissolve it. Now add a sufficient quantity, as explained below, of a solution of silver nitrate, which will usually cause a white precipitate of silver chloride, and then cautiously add ammonia until, if arsenic be present, a reddish precipitate of silver arsenate appears. If too

* This method was originally developed by Dr. Pearce and the author in the laboratory of the Boston and Colorado Smelting Works, at Argo, Colorado.

much ammonia be added, the precipitate first formed will redissolve and may not be observed at all. In this case the bit of litmus paper in the liquid will show an alkaline reaction. Now cautiously add nitric acid (best dilute) until the red precipitate just redissolves or the litmus paper shows a slight acid reaction. To the faintly acid liquid add a few cubic centimeters of a strong solution of sodium acetate or 1 or 2 grams of the crystals. This will effect a replacement of the free nitric acid with acetic acid and all the arsenic will be at once precipitated as silver arsenate, Ag_3AsO_4.

In order to avoid an unnecessarily large excess of silver nitrate, it is best to make up a solution containing, say, 17 grams in 500 cc. and use a definite amount. 1 cc. of a solution of this strength will precipitate 0.005 gram of arsenic, or 1 per cent. if 0.5 gram of ore is taken for assay. Thus 10 cc. will be an excess in most cases.*

Heat the precipitated mixture to boiling and then cool to room temperature, allowing the precipitate to settle somewhat, and filter. If the first portions run through turbid, return them to the filter once more. Test the filtrate with a little more silver nitrate and sodium acetate. Wash the precipitate with cold water until a portion of the washings shows only a faint cloud when tested for silver with a soluble chloride.

Now place the original flask under the funnel and dissolve the arsenate on the filter with cold dilute (1:1) nitric acid; 5 or 10 cc. will usually suffice. Wash the filter thoroughly with cold water. A white residue of silver chloride usually remains undissolved. Dilute the filtrate, if necessary, to about 100 cc., add about 5 cc. of a strong solution of ammonio-ferric alum, and titrate to a permanent red tinge with a solution of ammonium thiocyanate (Volhard), shaking well, especially at the end, to break up the clots of precipitate and free any solution held mechanically.

* It is best to use at least 10 cc. in every case, as small amounts of arsenic may entirely fail to precipitate unless a considerable excess of silver nitrate be added.

Multiply the number of cubic centimeters required, by the arsenic value of 1 cc. to obtain the amount of arsenic in the ore.

7. If the thiocyanate solution contains 7.612 grams of NH_4SCN per liter, 1 cc. will equal 0.0025 gram, or 0.5 per cent., of arsenic. It should be standardized carefully, either against pure silver or silver nitrate, which contains 63.51 per cent. of silver. Weigh accurately about 0.5 gram of pure silver, dissolve in a little nitric acid, and dilute to about 100 cc., or, instead, weigh about 0.8 gram of silver nitrate and dissolve in 100 cc. of water acidified slightly with nitric acid. In either case add a few cubic centimeters of a strong solution of ammonio-ferric alum (acidified with a little nitric acid) as an indicator, and titrate to a faint-red tinge with the thiocyanate solution as described above. From the number of cubic centimeters required and the weight of silver taken, determine the value of 1 cc. in silver. The formula of the red precipitate, Ag_3AsO_4, shows that 107.93 parts of silver represent 25 parts of arsenic, and, accordingly, the arsenic value of 1 cc. of the thiocyanate solution is easily deduced from the silver value by proportion.

8. When testing heavy sulphide ores there is some danger of a loss of arsenic by volatilization during the deflagration. To avoid this proceed as follows: Cover the dish, containing the weighed ore, with a watch-glass. Add a little nitric acid and then heat gently, with the cover on, to complete dryness. The dish may be placed above a small free flame and the heat increased at the end. Allow to cool, loosen any spatterings on the cover with a moistened rubber-tipped glass rod, and rinse them into the dish with as little water as possible. It is usually unnecessary to evaporate off this water if the amount is small. Add the 5 grams of nitrate mixture, heat cautiously to dryness, and then fuse and proceed as usual.

9. Sodium Carbonate and Zinc Oxide Method for Ores, etc.* —Thoroughly mix 1 part of dry sodium carbonate and 4 parts of zinc oxide. Weigh 0.5 gram of the ore into a platinum dish and intimately mix it with 3 grams of the above mixture, then cover with 2 grams more. Partially cover the dish with a thin piece of asbestos board, so as to keep in the heat, and heat over a Bunsen burner to bright redness for about twenty minutes. An uncovered porcelain dish may be used, similarly heated in a muffle. Allow to cool and transfer the mass to a suitable beaker. rinsing out the dish with hot water and making up the bulk in the beaker to about 50 cc. Heat the mixture to boiling, stirring well, and filter, washing thoroughly with hot water. Receive the filtrate in an 8-oz. flask. Drop a small bit of litmus paper into the solution as an indicator and then make slightly acid with acetic acid. Now add an excess of silver nitrate solution, as described in 6, and agitate the mixture occasionally for a few minutes without heating. Filter the silver arsenate, refiltering if any runs through, and finish the determination as described in 6.

Note.—For an arsenic determination, where antimony is not also required, this method is shorter and simpler than either of those previously described, but I have found it inapplicable to oxidized ores containing much lead, as some of the arsenic always combines with the lead and fails to be extracted. A determination can easily be made in about fifty minutes. The following test indicates that antimony does not interfere:

An oxidized ore treated by my own methods gave

Arsenic....................	9.365 per cent.
Antimony	0.47 "

* W. C. Ebaugh and C. B. Sprague, Jour. Am. Chem. Soc., XXIX, 1475.

In trying the above method with this ore, 0.5 gram was weighed into a platinum dish and 0.1 gram of Sb_2O_3 added. This was equivalent to increasing the antimony contents of the ore to about 17 per cent. Two grams of the zinc oxide mixture were intimately mixed with the above and about a gram more was used as a cover. The dish and contents were kept at a bright red heat over a Bunsen burner for 15 minutes and the assay then finished as described. Found, arsenic, 9.365 per cent, exactly checking my result by the other method.

10. Method of L. L. Krieckhaus.*—Decompose 0.5 gram of the ore with the zinc oxide mixture precisely as described in **9.** Receive the filtrate in an 8-oz. flask and boil it down to 50 cc. After cooling, add 50 cc. of strong hydrochloric acid and again cool. Now add 10 cc. of a 20 per cent potassium iodide solution (the solid salt would fail to properly dissolve in the strong acid) and allow a minute for the reaction to become complete, as follows:

$$H_3AsO_4 + 2HI = H_3AsO_3 + H_2O + I_2.$$

This reaction can be made to go completely in either direction according to the conditions.† In the strongly acid solution prescribed it proceeds as written. Now add about 100 cc. of cold water and titrate to complete decolorization with a standard solution of sodium thiosulphate. No indicator is required, as the end-point is easily observed. The thiosulphate solution used in the iodide method for copper (XIII, 1) may be employed. The copper value multiplied by 0.59 will give the arsenic value.

When the solution of the sintered mass, after the decomposition, is colored green or pink by manganese, add 5 cc. of

* Eng. and Min. Jour., Vol. 90, 357.

† E. W. Washburn, Jour. Am. Chem. Soc., XXX, 31.

alcohol or hydrogen peroxide and boil to precipitate the manganese before filtration. Chromium, if present, interferes as in the Pearce method. Phosphorus does not.

I have found this an excellent method. The treatment of the filtrate is simpler than the procedure of Sprague and Ebaugh and the result quite as accurate. There is the same objection to both methods, however, and that is the difficulty of decomposing oxidized lead ores so as to form a soluble alkali arsenate. Acting on Krieckhaus' idea, I have devised the following method, which appears to overcome this objection and be suitable in all cases, although I have not had the opportunity, as yet, of making many comparative tests.

11. Author's Zinc Oxide Method.—Prepare a mixture of 1 part sodium carbonate, 1 part potassium nitrate, and 2 parts zinc oxide. To 0.5 gram of the ore in a platinum dish add 5 grams of this mixture and grind intimately together with an agate pestle. Heat the dish and contents gradually to dull redness, then to bright redness for perhaps ten minutes. If a Bunsen burner is used, nearly cover the dish with a piece of thin asbestos board to prevent radiation. A porcelain crucible and muffle heating may be employed if desired. Cool, disintegrate the sintered mass with hot water and filter, washing the residue on the filter six or seven times with hot water. Receive the filtrate in an 8-oz. flask. Add to the filtrate 6 cc. of strong sulphuric acid (this should be an ample excess) and boil to strong sulphuric acid fumes. The final heating is best done over a free flame. Allow to cool, add 50 cc. of water, heat, if necessary, to effect solution, cool again, and add 50 cc. of strong hydrochloric acid. Once more cool the slightly warm solution, add 4 cc. of a 50 per cent. solution of potassium iodide (this is the usual solution for the iodide method for copper), mix and allow to stand about a minute. Now add 100 cc. of cold water and titrate with sodium thiosulphate solution as in 10. It is best to run a blank and deduct for the drop or so of thiosulphate required. Chromium and phosphorus do not interfere, nor does any small amount of manganese that may pass in solution through the filter.

CHAPTER VII.

BARIUM.

1. **Method for Ores.**—Decompose 0.5 gram of the ore in a 4-oz. Erlenmeyer flask, or a covered beaker, by one of the methods described for INSOLUBLE RESIDUE (XXIV), according to its nature. When solution is as complete as possible add a few drops of strong sulphuric acid to make sure that all the barium will be rendered insoluble. Dilute the mixture with about 50 cc. of water, heat to boiling, and filter, washing with hot water. If the ore contains lead add about 5 grams of ammonium chloride before boiling and filtering, in order to retain it all in solution. Place the filter and insoluble residue containing the barium in a platinum dish or crucible and ignite to burn off the filter-paper. Mix the cold residue with 3-5 grams of mixed sodium and potassium carbonates and fuse for a short time to convert the barium to carbonate After cooling, heat the melt with water until disintegration is complete. If a platinum dish has been used it will probably hold sufficient water for the purpose; a platinum crucible may be placed in water in a beaker. Filter, washing with dilute ammonia water until sulphates are all removed. If desired, the filtrate may be tested from time to time by collecting a portion in a test-tube, slightly acidifying with hydrochloric acid, then adding a little barium chloride solution and warming. When no white precipitate of barium sulphate forms the washing is sufficiently complete.

Rinse the more or less impure barium carbonate on the filter as completely as possible into a small beaker. Some barium carbonate will still remain on the filter, and there may also be a little adhering in the dish or crucible used for the fusion. Dissolve the latter in 5 cc. of strong hydrochloric acid and then transfer this to the beaker containing the bulk of the carbonate, covering at once with a watch-glass to avoid loss by spattering. Warm the solution, wash off the cover and remove it and then pour the liquid through the filter last used, so as to dissolve whatever barium carbonate was left there. Wash beaker and filter with hot water and receive the filtrate in a large beaker. Dilute to about 300 cc., heat nearly to boiling, and add about 3 cc. of strong sulphuric acid diluted with sufficient water to prevent violent action. Cover the beaker and allow to stand, hot, for several hours, best over night, to insure the complete precipitation of the barium sulphate. Finally filter and wash well with hot water. The precipitate is very fine and will run through a loosely woven filter. The best Swedish or German filters usually give no trouble. It is safest to decant carefully through the filter without disturbing the precipitate in the bottom until most of the clear liquid is gone, then remove the beaker containing the filtrate and replace it with another, so that if the barium sulphate runs through there will be less liquid to refilter. The combined filtrate and washings should be allowed to stand, hot, for a while longer, to be certain that the precipitation is complete. The moist precipitate and filter may be ignited together over a Bunsen burner in a weighed platinum or porcelain crucible. The strong heat of a blast-lamp should not be employed. The carbonaceous matter of the filter may reduce some of the barium sulphate to sulphide, but ignition with free access of air will easily effect reoxidation. The ignited barium sulphate should be perfectly white. The weight of the

$BaSO_4$ multiplied by 0.657 will give the corresponding weight of BaO if the latter is required.

In rapid technical work it is usually sufficient to ignite the barium sulphate in a small clay "annealing-cup," and, when cold, shake and brush it from the cup to the scale-pan for weighing.

In smelter practice the barium is usually required to be reported as sulphate, and all the barium in the ore is considered as existing as sulphate.

2. **Short Method.**—The following method, though not so reliable, will frequently serve for technical purposes:

Having obtained the insoluble residue, including the barium sulphate as above, ignite it in a small platinum dish to burn off the filter-paper, and then, after cooling, add a few cubic centimeters each of strong hydrochloric and hydrofluoric acids, in the order named, and evaporate on the water-bath nearly or quite to dryness. It is best to repeat the operation to insure the complete expulsion of the silicic acid. Finally, take up in hydrochloric acid, dilute with hot water, and filter. Wash the residue well with hot water and then ignite and weigh as $BaSO_4$ as described above.

CHAPTER VIII.

BISMUTH.

1. **Method for Ores, etc.**—Treat 0.5 gram of the ore in an 8-oz. flask with 6–10 cc. of strong nitric acid and boil nearly to dryness. Add 5 cc. of strong hydrochloric acid, or more if necessary, and heat gently until solution is as complete as possible, and then add 10 cc. of strong sulphuric acid and expel the more volatile acids by boiling over a free flame until fuming strongly. Cool and add 25 cc. of water and boil gently a few minutes to insure solution of all the bismuth sulphate. Cool again, filter, and wash with dilute sulphuric acid (1-10). Do not allow to stand too long before filtering or some basic bismuth sulphate may separate. Dilute the filtrate somewhat and pass in a current of hydrogen sulphide to saturation. Bismuth, copper, arsenic, antimony, etc., are precipitated as sulphides. Filter, washing with weak hydrogen sulphide water. Rinse the precipitate as completely as possible into a beaker, add 3–4 grams of pure potassium cyanide, and warm gently for some time. Bismuth sulphide will remain undissolved, also cadmium sulphide if present and any lead that may not have been removed as sulphate. Filter through the same filter as before, in order to act upon the traces of sulphides that could not be washed into the beaker, and wash with hot water. Spread out the filter on a watch-glass and rinse off the sulphides into a beaker. To any adhering residue on the filter add a little dilute (1:2) nitric acid and warm until dissolved. Now rinse this in with the main portion, add a little strong nitric acid if necessary, warm the mixture until the bismuth is all in solution and the separated sulphur clean.

Dilute a little and then filter and wash thoroughly with 1:2 nitric acid. Dilute the filtrate contained in a large beaker to about 300 cc. and heat to boiling. Nearly neutralize the hot solution with dilute ammonia and precipitate the bismuth as basic chloride precisely as described for refined lead (6). Only one precipitation will be necessary. Collect the precipitate on a weighed filter, or a Gooch crucible, dry at 100° C., and weigh as BiOCl. Multiply the weight found by 0.8017 to obtain the weight of the bismuth.

2. Instead of precipitating the bismuth as basic chloride, it may, in the absence of lead and cadmium, be precipitated as basic carbonate as follows: Partially neutralize the filtrate from the solution of the bismuth sulphide in nitric acid with ammonia, but without producing any permanent precipitate, and then add a solution of ammonium carbonate in very slight excess. Heat nearly to boiling for some time, until the bismuth carbonate has settled well, and then filter and wash with hot water. Dry the precipitate and transfer it to a small weighed porcelain crucible, removing it from the paper as completely as possible Burn the latter carefully and add the ash to be precipitated in the crucible. Ignite the whole at a low red heat, cool, and weigh as Bi_2O_3 Multiply the weight found by 0.8966 to obtain the weight of the bismuth.

The oxidation of the bismuth sulphide by nitric acid may produce some bismuth sulphate, which would cause a slight contamination of the basic bismuth carbonate with basic sulphate, and the latter would fail to be converted to oxide by ignition. The error thus introduced is ordinarily sufficiently small to be negligible in technical work. If, however, greater accuracy is desired, the bismuth may be reduced to metal by the method of H. Rose, as follows:

3. To the ignited oxide in the crucible add about five times

its weight of pure potassium cyanide and fuse the mixture over a Bunsen burner with the flame at about half the usual height. Under these conditions there is no danger of volatilizing bismuth, which boils at about 1600° C. The reduction is usually complete in about 20 minutes. Extract the cold melt with warm water so as to dissolve the salts and leave the metallic bismuth. The latter may be in one or more globules, either loose or adhering to the crucible, and usually contaminated with particles of the glaze. To prevent loss of these particles, filter the aqueous extract through a filter that has been dried at 100° C. and weighed with the empty crucible. After extracting and washing thoroughly with water, wash with absolute alcohol and ether and then place the filter again in the crucible, dry at 100° C., cool, and weigh. The gain in weight over the original weight of the filter and empty crucible represents the metallic bismuth.

4. **Electrolytic Method for Ores.** — The deposition of an adherent coating of bismuth on a cathode of the size herein described, and under the conditions named, requires that not much more than 0.03 gram of bismuth shall be present. This may be regulated either by the amount of ore taken for assay or by taking only a portion of the final solution of bismuth for electrolysis. The quantity of bismuth present in an ore may be roughly judged by the bulk of the BiS precipitate.

Take 0.5 gram of ore and proceed precisely as described above (1) until the separated bismuth sulphide has been dissolved in dilute nitric acid and the solution filtered, the filtrate being received in an 8-oz. flask. Now add 6–7 cc. of strong sulphuric acid and boil to white fumes over a free flame. Cool, dilute with 25 cc. of water, and boil gently until all the bismuth sulphate is dissolved. No appreciable amount of lead sulphate should be found. Cool, transfer to the proper beaker, and dilute to 100 cc. with cold water. The solution is now ready for electrolysis. The beaker and electrodes may be the same as used for copper, as follows. (See chapter on ELECTROLYSIS.)

Beaker. Diameter, about 5 cm.; height, about 8–9 cm. Have a mark at the 100-cc. point.

Electrodes. The cathode is a platinum cylinder 5 cm. long and 2.5 cm. in diameter. This gives a total surface of about 78.5 sq. cm. With a current of 0.6 ampere, $ND_{100}=0.76$. The weight of this electrode is about 12.5 grams.

The anode is a stout platinum wire with a flat spiral base, the straight portion of the wire rising out of the center of the base at right angles. Its weight is about 8.5 grams. In use the anode is placed within the cathode with the base a little below the lower edge of the cylinder.

Arrange the weighed electrodes in the beaker so that the top of the cylinder projects a little above the surface of the liquid and cover the beaker with the two halves of a split watch-glass. Electrolyze with a current of 0.6–0.7 ampere; electrode tension about 2.7–3 volts. In 1½ hours remove and weigh the cathode. To do this, disconnect and lift it from the beaker, at the same time washing it with a gentle stream of water, and then immerse it in a beaker of water. Lift it from this, wash off the water with alcohol, dry at 100° C., cool, and weigh. Now dissolve off the bismuth with nitric acid and ignite and weigh once more, and then replace in beaker and electrolyze for another half hour, when both electrodes may be removed and weighed. A slight deposit of Bi_2O_5 is usually found on the anode. Calculate this to Bi and add the weight to that of the cathode deposit. It is well to again clean and replace the electrodes and electrolyze for a while longer to be sure of obtaining the last traces of bismuth. Finally, test the exhausted solution with hydrogen sulphide water.

5. In case the bulk of the bismuth sulphide indicates more than 6 per cent. bismuth in the ore, proceed as follows: Dilute

the filtered nitric acid solution of the bismuth sulphide to 250 cc., adding more acid if a basic salt begins to separate, and then continue the analysis with a suitable aliquot portion. If the percentage of bismuth is approximately known the proper amount of ore may be taken at the outset.

6. **Bismuth in Refined Lead.***—Dissolve 25 grams of the lead, best rolled or hammered out thin and cut into small pieces, in a mixture of 250 cc. of water and 40 cc. of nitric acid of 1.42 sp. gr., using a large covered beaker. Warm gently until all the lead is dissolved, then remove the beaker from the heat and to the hot solution add dilute ammonia ($\frac{1}{3}$ strong ammonia and $\frac{2}{3}$ water) very cautiously, finally drop by drop, until the free acid is neutralized and the liquid remains faintly opalescent. There should not be a visible precipitate but just a faint cloudiness. Now add 1 cc. of dilute hydrochloric acid, 1:3. The solution will clear for an instant and then, if any considerable amount of bismuth is present, a crystalline precipitate of bismuth oxychloride will form. Again place the beaker over the heat so the liquid will keep hot but not boil. In an hour the bismuth oxychloride, together with a little lead, will have settled. Filter off the precipitate and wash it once or twice with boiling water. In addition to bismuth and lead, the precipitate may contain some antimony, if any appreciable quantity of the latter is present in the sample. Dissolve the precipitate on the filter in a small quantity of hot dilute hydrochloric acid 1:3, wash the filter with hot water and dilute the filtrate with water, taking care not to make the liquid so dilute as to cause a precipitation of the bismuth as basic chloride. Pass hydrogen sulphide into the liquid to precipitate all the bismuth, lead, and antimony as sulphides, filter, wash once with water and twice with warm ammonium sulphide to dissolve the antimony sulphide, and

* Ledoux & Co.

then, after washing once more with water, dissolve the precipitate of bismuth and lead sulphides by placing filter and contents in a small beaker and heating with dilute nitric acid (1:4). Boil so as to thoroughly disintegrate the filter-paper and then dilute somewhat with hot water and filter. Wash the filter well, first with a little warm dilute nitric acid (1:4) and then with hot water. Partially neutralize the nitric acid in the filtrate with ammonia, dilute with warm water to a volume of about 300 cc., and then complete the neutralization and add 1 cc. of dilute hydrochloric acid as described above for the original precipitation. The bismuth will now come down as basic chloride free from lead. Filter on a weighed filter or Gooch crucible, wash well with hot water, dry at 100° C., and weigh as BiOCl. This weight multiplied by 0.8018 will give that of the bismuth.

7. In the absence of appreciable amounts of antimony the precipitation by hydrogen sulphide and subsequent washing with ammonium sulphide may be omitted, as only a small portion of whatever antimony is present is precipitated with the bismuth. In this case the first precipitate, containing the bismuth and a little lead, may be at once redissolved on the filter with warm dilute nitric acid (1:4) and the precipitation repeated on the filtrate as described. Copper, silver, arsenic, and such amounts of iron as occur in refined lead do not interfere

8. **Bismuth in Lead "Bullion."**—The determination of bismuth in impure lead or lead bullion may be carried out on the same lines as described for refined lead. From 10 to 25 grams may be taken, according to the presumed or known nature of the material. Antimony may usually be assumed to be present. Proceed with the determination precisely as described for refined lead (6) up to the point where the bismuth, lead, and antimony sulphides are dissolved in hot dilute nitric acid. Now, instead of finishing the analysis as described, it is best to repeat

the whole series of operations once more. In this way the impurities are more certain to be removed and the BiOCl weighed pure.

9. Eakins's Method for Bismuth in Refined Lead.*—In this method for the determination of bismuth in refined lead and lead bullion the bulk of the lead is removed at the outset by precipitation as sulphate from the nitric acid solution. In my laboratory this method has uniformly given low results as compared to those obtained by the method of Ledoux & Co., some of the bismuth apparently being precipitated with the lead. For a description of the method reference is therefore made to the original paper, or to Furman's Manual of Practical Assaying, seventh edition, p. 193.

10. Volumetric Method for Ores.—The following scheme is based on Miller and Frank's modification of Riderer's method:†

Weigh 0.5 gram of the ore and proceed as in (1) until the washed sulphides of bismuth, etc., are obtained on the filter. Place filter and precipitate in an 8-oz. flask and heat with 5–10 cc. of 1:1 nitric acid until the separated sulphur is clean and the filter well disintegrated. Dilute a little and then filter and wash filter thoroughly with 1:2 nitric acid, best with the aid of a filter-pump. To the filtrate add 30 cc. of a cold saturated solution of tartaric acid, then drop in a bit of litmus paper as an indicator and make the solution slightly alkaline with potassium hydroxide. Now add sufficient potassium cyanide solution to dissolve any precipitate that may have formed (except possibly a little bismuth sulphide thrown down by sulphur in the cyanide), and then pass in hydrogen sulphide gas to saturation. Bismuth sulphide is precipitated; copper, arsenic, antimony, etc., remain in solu-

* Proceedings Colorado Scientific Society, Feb. 1895.

† Jour. Am. Chem. Soc., XXV, 926.

tion. Filter and wash with cold water. Dissolve the bismuth sulphide in dilute nitric acid and filter the solution precisely as described above.

11. The filtrate now contains the bismuth as nitrate together with some free nitric acid. It is best to have the excess of the latter about 5 per cent. Now add a decided excess (3 or 4 times the amount theoretically necessary for combining with all the bismuth) of the ordinary molybdic acid solution used for phosphorus determinations (XXII, 7). There should be no precipitate produced at this point. Then add a few drops of Congo-red solution (made by dissolving 1 part of Congo-red in 100 parts of 30 per cent. alcohol) and slowly run in, with stirring, very dilute ammonia from a burette. A white precipitate will form, and finally the indicator will become pink. Next add a few drops of dilute nitric acid, sufficient to change the color to lilac (just neutral). Now dilute, if necessary, to about 200 cc. and heat slowly on a thick asbestos pad, very hot (60° C.), but not to boiling, and stir vigorously. The precipitate of bismuth ammonium molybdate ($BiNH_4(MoO_4)_2$) should coagulate and appear like silver chloride, perhaps colored pink by the indicator. It should not be yellowish. Filter the hot mixture and wash the precipitate thoroughly with a 3 per cent. solution of ammonium sulphate. If the above directions are followed, the precipitation will be complete, but if too much nitric acid is added, so that the indicator is turned back to a decided blue color, the precipitate does not collect or filter as well and the filtrate may contain traces of bismuth. If the precipitate has a yellow color, the results will be unreliable. In such a case, make alkaline with ammonia, then add nitric acid until the precipitate is dissolved, and repeat the neutralization and heating. It is usually unnecessary to add more ammonium molybdate.

12. Make a mixture of 75 cc. of water and 15 cc. of strong

sulphuric acid. Place a portion of this in a beaker and rinse the precipitate into it from the filter as completely as possible. When all is dissolved pour the solution through the filter to dissolve any remaining precipitate and wash with the remainder of the acidulated water. Receive the filtrate in an 8-oz. flask, add 5 grams of 100-mesh granulated pure zinc, and allow to stand until the solution is of a green color (not brown) and the zinc has wholly or nearly dissolved. Now place a thick wad of absorbent cotton in a funnel, wet it and place a little granulated zinc on top. Filter the reduced solution through this and wash thoroughly with lukewarm water. Receive the filtrate in a somewhat larger flask, which has been filled with carbon dioxide by pouring a little dilute sulphuric acid over some sodium acid carbonate placed in the bottom. Warm the solution to about 40° C. and titrate with permanganate solution, the same that is used for iron determinations (XV, 2), as described in XXII, 9.

The formula $BiNH_4(NoO_4)_2$, indicates that $2Mo = Bi$, and it is shown in XX, 11, that $6Fe = 2Mo$; therefore $6Fe = Bi$, or, the iron factor of the permanganate multiplied by 0.6216 gives the bismuth factor.

CHAPTER IX.

CADMIUM.

1. **Method for Ores.**—To 0.5 gram of the ore in an 8-oz. flask add 10 cc. of strong hydrochloric acid and 5 cc. of strong nitric acid. Boil until any sulphides present are decomposed and the acid is perhaps half expelled. If oxides still remain, continue to heat with addition of hydrochloric acid until solution is as complete as possible. Finally, add about 6 cc. of strong sulphuric acid and boil, best over a free flame, until fuming copiously. Cool, add about 25 cc. of water, heat to boiling, and allow to stand, hot, for a short time, to insure the solution of anhydrous iron sulphate, etc. Then cool, filter off the insoluble residue (including any lead sulphate), and wash with cold dilute sulphuric acid (1:10) Dilute the filtrate to about 200 cc. and pass in hydrogen sulphide gas to saturation. Filter off the precipitated sulphides and wash with dilute hydrogen sulphide water slightly acidulated with hydrochloric acid. Rinse the precipitate from the filter into a beaker as completely as possible, using no more water than necessary, place the beaker under the funnel and pour through the filter a strong cold solution of pure potassium cyanide. Agitate the beaker so as to mix the liquids and use no more cyanide solution than necessary to dissolve the soluble copper sulphide, etc., that may be present. If no bismuth or lead is present (any lead should have been practically all removed as sulphate) the cadmium sulphide will now appear yellow or

orange.* Filter the mixture through the same filter as before. If it runs through turbid return it through the filter until clear. Wash well with dilute hydrogen sulphide water. The cadmium sulphide has a tendency to pack on the filter and impede filtration. If possible, wash it loose each time and the washing will then proceed much more rapidly. If the washed precipitate appears appreciably discolored, indicating bismuth, proceed as in 2, below. If clean and yellow or orange, dissolve it by pouring through the filter hot dilute (1:1) hydrochloric acid, using as little as possible. If the volume of the filtrate is not too large, receive it in a large weighed porcelain crucible; otherwise collect it in a beaker and transfer it to the crucible in small portions at a time. Place the crucible and contents on a water-bath and evaporate the solution of cadmium chloride to complete dryness. Now cover the crucible and add a slight excess of dilute sulphuric acid. When spattering has ceased, remove and rinse off the cover and continue the evaporation as far as possible on the water-bath. Finally, remove the excess of sulphuric acid by cautiously heating over a free flame until no more fumes are evolved. Avoid heating higher than necessary. It is best to place the crucible within a larger one fitted inside with an asbestos ring, so that the crucibles do not touch. Or, the crucible may be supported on a clay triangle on a small iron sand-bath containing no sand, so as to just not touch the bottom. When the fumes have ceased to come off, cool and weigh the crucible and contents. The cadmium sulphate, $CdSO_4$, should be pure white and soluble in water without residue. Its weight multiplied by 0.5391 will give the weight of the cadmium, from which the percentage may be calculated.

2. If the cadmium sulphide, after treatment with potassium

---

* When precipitated from sulphuric acid solutions, unless very weak in acid, the cadmium sulphide is orange-colored rather than pure yellow.

cyanide, appears dark-colored, bismuth or lead (possibly mercury) may be present. Place the moist precipitate and filter in an 8-oz. flask and add 10 cc. of strong hydrochloric acid and the same amount of water. Boil the mixture until the cadmium sulphide is all dissolved, the hydrogen sulphide expelled, and the filter well disintegrated. Any dark insoluble residue apparently free from cadmium may be neglected. Dilute with 25 cc. of hot water and filter, washing thoroughly with hot water. Dilute the filtrate somewhat, add sodium carbonate in slight excess and then 1 or 2 grams of potassium cyanide. Digest for some time at a gentle heat and filter. Wash with cold water. Bismuth and lead remain on the filter as carbonates. Pass hydrogen sulphide through the filtrate, diluted if necesary. This should precipitate pure yellow cadmium sulphide unless mercury is present, which is rarely the case. Filter, wash with hydrogen sulphide water, and then dissolve in hydrochloric acid and finish as described above.

3. **Electrolytic Method.**—Cadmium can be separated electrolytically in a satisfactory manner from its solution in various electrolytes. Its alkaline double-cyanide solution is perhaps as good as any, if not the best electrolyte. The method may be applied to an ore as follows:

4. Treat 1 gram of the ore by the methods previously described until a solution of the chloride or sulphate is obtained, free from the other members of the hydrogen sulphide group. The solution should be evaporated if necessary, so as to have the volume well below 100 cc. Add a drop or two of phenolphthalein solution and then pure sodium or potassium hydroxide solution until a permanent red color is obtained. Now add cautiously a strong solution of pure potassium cyanide until the precipitated cadmium hydroxide is completely dissolved, avoiding an excess. Dilute to 100–125 cc. and, using the same beaker

and electrodes as for copper (XIII, 10), electrolyze with a current of $ND_{100}=0.04$ to 0.06 A. and V. = 2.9–3.2. Keep the solution at a temperature of about 60° C. In from 4 to 6 hours the deposition is usually complete. Test by raising the level of the liquid slightly and noting if any deposition occurs on the clean platinum surface during half an hour or so. When the operation is ended, disconnect and remove the electrode, wash it with hot water, then with alcohol, and finally dry it at 100° C., cool, and weigh.

It is a good plan to replace the electrode in the solution and electrolyze for another half hour. The electrode should first be cleaned with nitric acid, then washed, ignited, and again weighed. If desired, the solution can be tested with hydrogen sulphide. As this may produce a yellow color in the cyanide solution, even in the absence of cadmium, it is best to first acidify with hydrochloric acid (under a hood) and then heat until all the hydrocyanic acid is expelled.

5. Treadwell,* using the same electrolyte, recommends electrolyzing in the cold for from 5 to 6 hours with a current of 0.5–0.7 ampere and an electromotive force of 4.8–5 volts. At the end of this time the current is increased to from 1–1.2 amperes and the solution is electrolyzed for 1 hour more. Treadwell states that unless the current is increased toward the end of the operation, the cadmium will not be all deposited at the end of 12 hours. If the stronger current is used from the beginning, some of the metal is liable to be deposited in a spongy form, resulting in a possible loss in washing.

6. Smith † describes the deposition of cadmium from its sulphate solution. The neutral solution is acidified with 3 cc. of sulphuric acid of 1.09 sp. gr. and diluted to 125 cc. Lily G.

* Treadwell's Analytical Chemistry (Hall), II, 2d Ed., p. 166.

† Electrochemical Analysis.

Kollock * found that with such a solution, maintained at 65° C., with ND_{100}=0.078 ampere and volts=2.61, the deposition was complete in 5 hours.†

* Jour. Am. Chem. Soc., XXI, 925.

† See Edgar F. Smith's Electro-Analysis for rapid method, using rotating electrode.

CHAPTER X.

CALCIUM.

1. Ores, etc.—The calcium is usually required as CaO. Treat 0.5 gram of the ore in an 8-oz. flask with whatever acids are best suited to decompose it. Begin ordinarily with 10 cc. of strong hydrochloric acid and heat gently. If sulphides are present, add 5 cc. of strong nitric acid (after the oxides are dissolved) and boil gently until decomposed. If the decomposition is perfect sulphuric acid is unnecessary, but with ores I usually add 5 cc. at this stage. In any case boil until most of the free acid is expelled, over a free flame if sulphuric acid has been used. Now cool, if necessary, add about 100 cc. of water and 5 cc. of strong hydrochloric acid, heat until the salts are dissolved and then filter (never omit this if there is any appreciable amount of insoluble residue), washing with hot water. To the filtrate add 2 cc. of a saturated solution of potassium dichromate (this may be omitted if lead is absent*), 15–20 cc. of saturated bromine water (to precipitate manganese), and then a very slight excess of ammonia. Cover the beaker and boil until all free ammonia is expelled. Allow to settle somewhat and then filter through an 11-cm. filter and wash thoroughly with hot, dilute ammonium chloride solution (5 grams to 500 cc.). If lead chromate runs through, return the first portions. If the precipitate shows much manganese it is always well to test the

* Lead unquestionably interferes, although but slightly. An ore containing 30 per cent. lead and about 16 per cent. CaO gave results about 0.25 per cent. high when the lead was not removed. Lead may also be removed as sulphide, as described in XVII, 1.

filtrate by adding more bromine water and ammonia and boiling a short time. If any manganese comes down, filter it off, it being usually unnecessary to expel the excess of ammonia as before. Wash with the dilute ammonium chloride solution.

By proceeding as above, the usual double precipitation of the iron is entirely unnecessary. The usual retention of lime by iron appears to be due to the formation of calcium carbonate, by the absorption of carbon dioxide from the air in the presence of free ammonia, and this does not occur if the ammonia is expelled previous to filtration. A neglect to filter off the insoluble residue before the iron precipitation will usually produce a low result. I am unable to state the reason.

2. Make the filtrate from the iron strongly ammoniacal, heat to boiling and add an excess of ammonium oxalate solution. This should be added in sufficient amount to convert all possible calcium and magnesium to oxalates (the magnesium oxalate remaining in solution). Thirty cc. of a saturated solution should be sufficient in any case. It is best to dilute it somewhat and add it boiling hot. Boil the mixture about 10 minutes, allow to stand hot and settle, and then filter through an 11-cm. filter and wash ten times with hot water.

The mode of procedure now depends upon whether the ore contains much or little magnesium. In the former case the calcium oxalate is almost sure to contain an appreciable amount of magnesium oxalate and should therefore be purified. Unless the amount of magnesium is known to be insignificant it is always safest to proceed as follows:

3. Without troubling to wash the precipitated oxalate more than once, rinse it from the filter into a beaker. What little remains adhering to the filter may usually be neglected if the same filter be employed for the next filtration of the oxalate.

Heat the mixture in the beaker and then dissolve the oxalates by the addition of as little hydrochloric acid as possible. Dilute to about 50 cc. with hot water, make alkaline with ammonia, and add about 10 cc. of the ammonium oxalate solution. Boil the mixture for about 10 minutes, allow to stand, hot, until settled, and then filter through the last filter. Wash at least ten times with hot water, to remove every trace of ammonium oxalate.

When little or no magnesium is present, the first precipitate of calcium oxalate will be sufficiently pure and require only thorough washing with hot water.*

4. In a 400-cc. beaker prepare a mixture of 5–7 cc. of strong sulphuric acid and about 125 cc. of water. Drop the filter and precipitate into this and heat to about 70° C. Stir to effect the decomposition of the calcium oxalate, but avoid disintegrating the filter. Titrate the hot solution with standard potassium permanganate solution. Under the conditions named the filter-paper will exercise practically no influence on the result, and while the final pink tinge will gradually fade, there will be no difficulty in noting a sharp end-point. Multiply the number of cubic centimeters of permanganate required, by the percentage value in CaO of 1 cc. to obtain the per cent. of CaO in the ore.

5. The permanganate solution used for iron determinations (xv, 2) will serve for this titration. The iron factor multiplied by 0.5017 will give the CaO factor. This is because the same amount of oxygen from the permanganate is required to oxidize the oxalic acid in the calcium oxalate corresponding to 1 mol. of CaO, as is required to oxidize 2Fe from the ferrous to the ferric condition. That is, 2Fe are equivalent to 1 CaO or 111.8 parts of Fe to 56.1 parts of CaO, which is the same as 1 part Fe to 0.5017 part CaO.

---

* This is the usual practice in all cases in ordinary lead-smelter work.

6. It is, however, best for the CaO determination to standardize the permanganate directly against pure oxalic acid. Weigh about 0.2 gram of the crystals and dissolve in an 8-oz. flask in a previously prepared mixture of 5–7 cc. of strong sulphuric acid and 125 cc. of water. Heat to about 70° C. and titrate to a faint pink tinge with the permanganate solution. The first portion of permanganate will not be decolorized instantly, but when once the decolorization has begun, it will thereafter occur quickly until the end.

Divide the weight of oxalic acid taken by the number of cubic centimeters required. This will give the oxalic acid value of 1 cc. of the permanganate. Multiply this figure by 0.4451 to obtain the value of 1 cc. in CaO, expressed in grams. This last factor is obtained as follows: 56.1 parts of CaO require 126.048 parts of crystallized oxalic acid to form calcium oxalate. Accordingly, 1 part of oxalic acid is equivalent to 0.4451 part of CaO. Of course, in the CaO determination all the oxalic acid titrated is derived from the decomposition of the calcium oxalate by the sulphuric acid.

Having determined the value of 1 cc. in grams of CaO, the percentage value on the basis of 0.5 gram of ore taken is easily calculated.

Example.—Took 0.2163 gram of oxalic acid. Used 38.66 cc. of permanganate. $0.2163 \div 38.66 = 0.005594$. This is the weight of oxalic acid to which 1 cc. of the permanganate is equal. Multiplying by 0.4451 we have 0.002491, the value of 1 cc. in CaO, or 1 cc. = 0.4982 per cent. on the basis of 0.5 gram of ore taken.

7. **Silicates and Substances Not Decomposed by Acids.**—These are decomposed, as described under SILICA, either by immediate fusion with alkali carbonate or by acid treatment, followed by fusion of the insoluble residue. The nature of the substance will determine these points precisely as in the case

of silica. When the solution eventually obtained contains considerable silicic acid, it is best to render it insoluble by evaporation and filter it off. A small amount of silicic acid may be neglected. The clear hydrochloric acid solution finally obtained containing all the calcium, is treated as described above (1).

8. **"Available Lime" in Ores containing Calcium Fluoride.**—Some of the western smelters regard the calcium contained in calcium fluoride as of no value as a flux, claiming that it passes through the furnaces unaltered, without combining with the fluxes used. In paying for the lime in an ore, only that which is not combined with fluorine is taken into consideration and is called "available lime." The method used for determining this portion is not strictly correct, since if an ore contained much calcium sulphate or silicate, at least some of the CaO thus combined would probably fail to be reported as "available." As a technical method, however, in regular use, it is sufficiently satisfactory and is therefore given herewith.

9. Take 0.5 gram of the ore in an 8-oz. flask, moisten with water, add 5 cc. of 80 per cent. acetic acid, and boil nearly to pastiness. Take up in about 30 cc. of equal parts of 80 per cent. acetic acid and water and boil gently for a few minutes. To remove lead, which is frequently present, add 2 cc. of a saturated solution of potassium dichromate just previous to this last boiling. Filter, washing with hot water. To the filtrate add about 20 cc. of saturated bromine water and then (cautiously) make alkaline with ammonia and heat to boiling. It is generally best to now add more bromine water and continue the heating for a short time to make sure of the precipitation of all manganese. Finally, boil off all the free ammonia, allow to settle somewhat, and filter and wash as in the regular method (1). Test the filtrate for unprecipitated manganese if apparently necessary, as described for the regular method. Make the final filtrate

strongly ammoniacal, heat to boiling, precipitate the calcium as oxalate, and finish the determination exactly as described for the regular method.

This will give all the CaO contained in the ore as carbonate and also as sulphate unless the latter is in large amount. Calcium combined as fluoride or silicate remains practically undissolved by the acetic acid. The silicate is usually small in amount and the calcium so combined is, in fact, regarded as no more "available" than that existing as fluoride.

10. The Percentage of Calcium Fluoride may be determined with fair accuracy as follows: Place about 3 grams of powdered anhydrous sodium sulphate in a small platinum dish, mixing in also a little potassium nitrate, if reducible metals are liable to be present. Lay the filter and residue from the above acetic acid treatment upon this mixture and ignite gently until the paper is fairly well burned off. Now cool and add 5-6 cc. of strong sulphuric acid. Heat carefully, to avoid spattering, first to strong fumes, and then to a melt, if possible. If the mass solidifies at the end, without melting, cool sufficiently, add a little more sulphuric acid and heat again. This will usually effect complete decomposition of the fluoride, even if the mass is not completely melted. Allow to cool, cover the dish and dissolve the cake by warming with sufficient water, acidulated with 5 cc. of hydrochloric acid. Transfer the solution to the original flask, first filtering, if there is an appreciable amount of insoluble residue. Dilute to about 150 cc. with hot water and proceed with the usual determination of CaO as described in 1, at the same point. Multiply the percentage of CaO found by 1.392 to obtain the percentage of CaF_2.

11. Rapid Volumetric Determination of CaO in Limestone, Cement, Lime, Blast-furnace Slags, etc.*—The following rapid

* From paper by Richard K. Meade, Chemical Engineer, Vol. 1, p. 20.

method for the determination of CaO is applicable to materials in which the calcium is present either as oxide, carbonate, or silicate, and is dependent on the fact that calcium can be completely precipitated as oxalate in solutions containing free oxalic acid, while iron, aluminum, and magnesium are not.

12. Decomposition of the Sample.—For high-grade *limestones*, that is, those which when burned do not give a hydraulic lime, weigh 0.5 gram into a platinum crucible cover and ignite for 5 minutes over a Bunsen burner, and then for 5 minutes over the blast-lamp. This heating must be cautiously carried out, as magnesian stones are likely to fly out if heat is applied too suddenly. Start the ignition over a low Bunsen flame and gradually raise until the full heat is attained, then continue for 5 minutes and follow with a blast-lamp. Empty the contents of the crucible into a 500-cc. beaker and add 40 cc. of dilute (1:1) hydrochloric acid; heat, and when solution of the sample is complete proceed as in **13**.

For *cement rock* or *hydraulic limestones*, weigh the sample as before and carefully mix with it $\frac{1}{4}$ gram of finely ground sodium carbonate by stirring with a glass rod. Brush off the rod into the crucible and ignite over a Bunsen burner, starting with a low flame and gradually raising it until the full heat is attained. Continue heating for 5 minutes longer and then ignite over the blast for the same length of time. Place the crucible in a 500-cc. beaker and decompose the sintered mass in the crucible with 40 cc. of dilute (1:4) hydrochloric acid, keeping the beaker covered to avoid loss by effervescence. Heat until solution is complete and proceed as in **13**.

For *cement*, pass the sample through a 100-mesh screen, weigh 0.5 gram into a dry 500-cc. beaker and add, with constant stirring, 20 cc. of water. Break up the lumps, and when all the sample is in suspension except the heavier particles, add 20 cc. of dilute

(1 : 1) hydrochloric acid and heat until solution is complete. This usually takes 5 or 10 minutes. Proceed as in 13.

Many *slags* are soluble in concentrated hydrochloric acid. When this is the case, weigh 0.5 gram into a 500-cc. beaker, stir up with a very little water and add 20 cc. of strong hydrochloric acid and heat. When solution is complete proceed as in 13.

13. **The Determination.** — Carefully add dilute ammonia (sp. gr. 0.96) to the solution of the sample until a slight permanent precipitate forms. Heat to boiling and add 10 cc. of a 10 per cent. solution of oxalic acid. Stir until the iron and aluminum hydroxides are entirely dissolved and only a slight precipitate of calcium oxalate remains. Now add 200 cc. of boiling water and sufficient (20 cc.) saturated solution of ammonium oxalate to precipitate the calcium. Boil and stir for a few moments, remove from the heat, allow the precipitate to settle, and filter on an 11-cm. filter. Wash the precipitate and paper 10 times with hot water, using not more than 10 or 15 cc. of water each time. Remove the filter from the funnel, open and lay against the sides of the beaker in which the precipitation was made, wash from the paper into the beaker with hot water, add dilute sulphuric acid (4), fold the paper over and allow to remain against the walls of the beaker. Heat to 80° C. and titrate with standard potassium permanganate (5) until a pink color is obtained; now drop in the filter-paper, stir until the color is discharged, and finish the titration carefully, drop by drop.

14. The permanganate is best standardized for this determination by means of pure calcite or Iceland spar, as this does away with uncertain factors and also with the error introduced by the solubility of calcium oxalate. The procedure, which is as follows, is that recommended by the Committee of the Lehigh Valley Section of the American Chemical Society:

Make up potassium permanganate solution by taking 6 grams of the salt to 1 liter of water; let stand a few days before standardizing.

Weigh out 0.5 gram of powdered calcite into a 400-cc. beaker; add 100 cc. of water and 10 cc. of hydrochloric acid (1 : 1); boil gently until all carbon dioxide is expelled, and when completely dissolved make alkaline with ammonia and add, little by little, 20 cc. of boiling-hot saturated solution of ammonium oxalate; continue boiling for 5 minutes; let settle, filter, wash, transfer precipitate to original beaker (13); dissolve in dilute sulphuric acid and titrate with the permanganate as in the determination of calcium.

15. Notes on the Foregoing Method.—In this method lead appears to interfere more or less, bringing the results high; it is therefore not so suitable for ores in general as the longer method first given.

In neutralizing with ammonia it is best to neutralize nearly all the free and combined acid, and the ammonium oxalate added later should be in sufficient excess to complete this neutralization of the mineral acid.

CHAPTER XI.

CHLORINE.

1. **Mohr's Volumetric Method.**—The chloride solution should be cold and neutral. If acid, it should be neutralized with pure sodium or calcium carbonate in slight excess. To the cold neutral, or faintly alkaline solution contained in a porcelain casserole or evaporating-dish add 1 cc. of a 2 per cent. solution of neutral potassium chromate. Titrate with a N/10 solution of silver nitrate until a permanent faint-red tinge is obtained, due to the formation of silver chromate. This compound cannot exist permanently in the mixture until all the chlorine has been precipitated as silver chloride. The mixture should be well stirred after each addition of silver nitrate, which toward the last should be added only drop by drop. As the faint reddish tinge is somewhat difficult to distinguish, various schemes have been proposed to facilitate its detection. I have found it a good plan, when the end-point is apparently attained, after reading the burette, to pour off half the liquid into a similar casserole and then add 1 more drop of the silver nitrate solution to one casserole and note if the two portions of the liquid now show any difference. When such a difference can be detected, the end-point has certainly been reached and it is usually safe to accept the reading of the burette taken previous to the last drop. For accurate work a blank test should be made on the same

volume of liquid to see how much silver solution is required to produce a tint when no chloride is present, and this amount must be deducted from that used in the analysis. Multiply the number of cubic centimeters used by 0.003545 to obtain the weight of the chlorine present, or by 0.00585 for the weight of the corresponding sodium chloride.

2. The standard N/10 silver nitrate solution for the above titration may be prepared as follows: Heat pure silver nitrate to 120° C. for 10 minutes, then cool and weigh 16.994 grams, dissolve in water and dilute to 1 liter. 1 cc. = 0.003545 gram of chlorine, or 0.00585 gram of sodium chloride. Of course any other weight of silver nitrate can be taken and the corresponding value of the solution obtained by calculation.

3. **Chlorine in Salts and Liquids Containing Organic Matter and Sulphides.**—*Solutions Required:* 1/10 normal silver nitrate. Approximately 1/10 normal ammonium thiocyanate. For ammonium thiocyanate take 8 grams per liter; or, for potassium thiocyanate, take 10 grams per liter. Standardize the thiocyanate against the silver solution, as described below, to obtain the value of 1 cc. in silver solution.

Indicator solution of ferric ammonium sulphate. Make a strong solution and add sufficient nitric acid to remove most of the brown color.

Procedure: Take 200 mg. of the salt, or an equivalent amount of the liquid. Place in an 8-oz. flask and dilute to about 50–75 cc. Add dilute nitric acid in moderate excess and boil a minute or two. Now run in the silver solution from a burette until sure of a slight excess and then boil the mixture until the silver chloride settles well on short standing. Cool completely, add a few cc. of the ferric indicator and titrate with the thiocyanate solution to a faint brownish tint. Note the amount of thiocyanate used, calculate its value in silver solution and deduct this from the amount of silver solution used. The remaining silver solution is the amount required to form silver chloride with the chlorine in the salt or liquid taken, and from this all necessary calculations can be made, as in 1.

In standardizing the thiocyanate, simply run 25–30 cc. of the silver solution into a flask, dilute to about 75 cc., add some ferric indicator and titrate in the cold to a faint color, shaking well, especially toward the end, after each addition of thiocyanate.

Precipitated AgCl does not interfere with the thiocyanate titration if it has been well coagulated by boiling. If allowed to remain in an opalescent or cloudy condition it will quickly decompose the slight excess of thiocyanate that causes the end-reaction color. The color will disappear repeatedly as more thiocyanate is added and prevent a correct end-point. This is because silver thiocyanate is more insoluble than silver chloride.

CHAPTER XII.

CHROMIUM.

METHODS are given below for the determination of chromium in iron ores, chrome-iron ore, and steel. These will include most of the cases likely to be met by the metallurgical chemist.

1. **Method for Iron Ores with Small Amounts of Chromium.*** —Fuse 1 gram of the very finely ground ore with a mixture of 5 grams of sodium carbonate and 0.5 gram of potassium nitrate in a platinum crucible or small dish. After fusion, extract the melt with hot water and transfer the mixture to a small beaker. If the solution is colored by manganese, add a little alcohol and warm the mixture. This will precipitate the manganese as dioxide. Allow the precipitate to settle and note the color of the clear solution. If chromium is present it will be more or less yellow. If quite colorless, chromium may be considered absent. Filter the mixture, washing with water, and dry and ignite the insoluble residue on the filter. Now grind it with ten times its weight of sodium carbonate and a little potassium nitrate, fuse the mixture and extract with water, etc., as before. Filter and add the filtrate to the former one. Acidify the combined filtrates with hydrochloric acid and evaporate to dryness to render the silica insoluble and reduce the chromic acid to Cr_2O_3. Take up in hydrochloric acid, dilute, and filter. Precipitate the Cr_2O_3 and Al_2O_3 in the filtrate with ammonia. Boil for a short time, filter, and wash well with hot water. Dry and ignite the precipitate and then fuse it with as little sodium carbonate and potassium nitrate as possible. Extract the melt with water and transfer the mixture to a platinum dish. Evap-

* Mainly from Blair. Chemical Anal, of Iron.

orate the liquid until it is very concentrated, adding crystals of ammonium nitrate from time to time to change all the carbonated and caustic alkali to nitrate. Each addition of the ammonium nitrate produces an effervescence and ammonium carbonate is given off. The solution finally becomes almost syrupy and smells faintly of ammonia, the addition of ammonium nitrate no longer causing an effervescence. Now add a few drops of ammonia and filter from the precipitated alumina, aluminum phosphate, manganese dioxide, etc. The filtrate contains the chromium as alkali chromate. Add an excess of a strong solution of sulphur dioxide, which changes the color of the solution from yellow to green. Boil well, add an excess of ammonia, boil again for a few minutes, filter on an ashless filter, and wash thoroughly with hot water. Dry and ignite the precipitate and weigh as Cr_2O_3. Multiply the weight by 0.6846 to obtain that of the chromium.

2. **Method for Chrome Iron Ore.**—Grind the ore in an agate mortar to the finest possible powder. Weigh 0.5 gram into a nickel or copper crucible of about 20-cc. capacity, in which have previously been placed 3 or 4 grams of sodium peroxide. In taking the sodium peroxide from the bottle or can, reject any white crust on top, which consists mainly of carbonate and oxide, and select only the yellow material. Thoroughly mix the contents of the crucible, and then heat gently over a Bunsen burner turned very low until the mass is entirely liquid. This may take about 10 minutes. Keep in a condition of low redness for about 10 minutes. Allow to cool until a crust forms on top and then add 1 gram more of sodium peroxide and fuse again at low redness for about 5 minutes. Cool, place the crucible and contents in a porcelain dish, add 50–100 cc. of water, and heat for a few minutes until the mass is dissolved. Remove and rinse off the crucible. If the solution is purple add a little more sodium

peroxide. Boil the covered solution for 10 minutes to decompose the sodium peroxide, since if any of the latter were allowed to remain it would, when the solution was subsequently acidified, react on the chromate and reduce the chromium to Cr_2O_3. The solution is now so strong in sodium hydroxide that it would probably rot a filter unless very largely diluted. To avoid the dilution the sodium hydroxide may be partially changed to carbonate by the addition of ammonium carbonate, but this must not be done until the sodium peroxide has been decomposed by boiling, as otherwise some ammonium nitrite would be formed which would cause a low result. Therefore, after the boiling, add 5 grams of ammonium carbonate, which should be sufficient to neutralize about four-fifths of the sodium hydroxide present, still leaving an excess of the latter. As soon as the ammonium carbonate has dissolved the liquid is ready for filtration. Filter off the insoluble material and wash it thoroughly, receiving the filtrate in a large beaker. Treat the residue remaining on the filter with hydrochloric acid. If any of it proves to be insoluble it may be undecomposed ore and it should be fused again with sodium peroxide.

3. Acidify the filtrate with dilute sulphuric acid (1:4) and then add a considerable excess—25 cc. or more. Allow the solution to cool and then transfer to a battery-jar (such as that used for iron titrations, XV, 7) and dilute to 700 cc. with cold water. Now add a weighed amount of ferrous ammonium sulphate which is more than sufficient to reduce the chromium present. To do this, place a sufficient quantity of the salt in a weighing-bottle and carefully weigh the whole; then add the salt in small portions to the chromium solution, while stirring, until the yellow color of the chromic acid has entirely disappeared, and also until a little permanganate run in is at once decolorized. The permanganate thus used counts as part of the total.

Now, without delay, titrate the excess of ferrous iron salt with standard potassium permanganate, and then again weigh the bottle of ferrous ammonium sulphate to determine the amount used. The weight of ferrous ammonium sulphate consumed in reducing the chromic acid is thus found, from which the corresponding weight of the chromium or Cr_2O_3 may be calculated.

4. The regular permanganate solution used for iron may be employed for the titration. It is best to standardize it for this titration directly against ferrous ammonium sulphate. Weigh out portions of about 1.5 grams each of the salt and dissolve in a battery-jar in 700 cc. of cold water containing 10 cc. of strong sulphuric acid. Titrate at once to a pink tint, or, better, to the first perceptible change of tint. Also run a blank to determine the correction required for the water and acid (see XV, 11).

The value of 1 cc. of the permanganate in ferrous ammonium sulphate is thus found.

167.7 parts of ferrous iron are required to reduce 52.1 parts of chromium in chromic acid to Cr_2O_3, or 1 part Fe=0.3107 parts Cr. The ferrous ammonium sulphate contains 14.25 per cent. of ferrous iron; therefore 1 part of the salt=0.04427 part of Cr.

Having found by the titration the excess of ferrous ammonium sulphate and deducted this from the total amount of the salt used, the weight of the remainder multiplied by 0.04427 will give the weight of the chromium sought, or, if multiplied by 0.06468, the weight of the Cr_2O_3.

5. **Method for Steel.***—This method is based upon the well-known fact that chromic salts can be oxidized completely to chromic acid by the addition of potassium chlorate to a concentrated nitric acid solution, and the fact also that the presence of

* From paper read by A. G. McKenna before the Chemical Section of the Engineers' Society of Western Pennsylvnia, June 18, 1896.

nitric acid does not interfere with the titration of chromic acid in a cold solution by means of ferrous sulphate and permanganate.

Weigh 3 grams of steel into a 400-cc. flask, add 35 cc. of strong hydrochloric acid, and boil for 5 or 10 minutes, which will be found sufficient to dissolve completely even the highest chrome-steels. When most of the hydrochloric acid is boiled off, add 150 cc. of strong nitric acid and continue the boiling until no more brown fumes are seen at the mouth of the flask, showing that the hydrochloric acid has all been driven off. Remove the flask from the flame or hot plate, allow to cool for 2 or 3 minutes, and then add 10 grams of potassium chlorate in crystals. It is best to allow the solution to cool somewhat before adding the chlorate in order to diminish the violence of the effervescence due to the action of the chlorate on the chromic salts. Replace on the hot plate and boil down to about 40 cc. in order to completely decompose the potassium chlorate. It is necessary to decompose the chlorate completely or results will be from 0.1 to 0.2 per cent high, but the amount of nitric acid left in the solution is unimportant. At this stage the chromium will all be in the solution in the form of chromic acid. Any manganese will be precipitated as dioxide, and generally some crystals of potassium nitrate arising from the decomposition of the chlorate will have separated out. Add 100 cc. of water and a few drops of hydrochloric acid. This will at once dissolve the manganese dioxide without action on the chromic acid. Boil the solution for a few minutes to remove the chlorine set free by the reduction of the manganese dioxide and then cool to room temperature. Make the cold solution, contained in a battery-jar, up to 700 cc. with cold water and add a weighed excess of ferrous ammonium sulphate, as described in 3, above. Finally, titrate the excess of ferrous salt with permanganate and calculate the result as in 4.

In very many chrome-steels the amount of manganese is so inappreciable in comparison with the chromium that for practical results it is not necessary to dissolve the dioxide as described above, but the solution after the evaporation to 40 cc. may be diluted and titrated at once.

5. **Rapid Method for the Determination of Chromium in Chrome and Chrome-nickel Steel.***—Dissolve 1 gram of the steel by warming in a large covered beaker with 25 cc. of the acid mixture described below. When solution is complete remove from the heat and add about 15 or 20 cc. of cold water. Rotate the liquid in the beaker and drop in at once about 1 gram of sodium bismuthate and continue to rotate the mixture for a few seconds. Now heat to boiling and boil rapidly. The permanganate formed by the sodium bismuthate will be rapidly decomposed and there will remain a clear violet liquid (manganic metaphosphate). Further boiling will complete the conversion of the chromium to CrO_3.

Decompose the excess of manganic metaphosphate with one-half cc., or more if necessary, of dilute hydrochloric acid and boil one minute. Cool somewhat, dilute to 200 cc. with cold water, add a slight excess of ferrous ammonium sulphate (as in 3) and titrate with permanganate. Calculate the result as in 4.

The end-reaction is very sharp, the results are accurate and an analysis can be completed in 10 minutes. The method has not yet been tested in presence of tungsten and molybdenum.

Acid mixture for dissolving the steel:

300 cc. nitric acid, 1.42 sp. gr.
300 cc. sulphuric acid, 1 part acid, 3 parts water.
300 cc. water.
100 cc. phosphoric acid, 85 per cent.
1½ grams manganese sulphate.

---

* N. M. Randall, Mining Science, LXI, 56.

CHAPTER XIII.

COPPER.

(See Appendix.)

DURING the many years that the iodide method for copper has been used in my laboratory it has been constantly studied, and from time to time slightly modified, until it is now the most accurate practical method for ores with which I am acquainted.

While the accuracy of the electrolytic method cannot perhaps be exceeded, the electrolytic method as actually carried out in some laboratories is liable to give erroneous results, principally owing to the failure to remove interfering impurities found in many ores. The proper removal of these impurities is apt to be quite tedious and involve considerable manipulation, tending to cause loss. I therefore give the iodide method the preference in most cases as being more practical and nearly if not quite as accurate as the electrolytic at its best.

1. **Iodide Method.***—A standard solution of sodium thiosulphate is required. Make up a solution containing about 19.5 grams of the pure crystals to the liter. Standardize this as follows: Weigh carefully about 0.2 gram of pure copper-foil and place in an 8-oz. flask (Whitall Tatum Co., pear-shaped). Add, best from a small pipette, 5 cc. of strong nitric acid (sp. gr. 1.42) which will quickly dissolve the copper. Dilute the solution a little and boil to expel the red fumes. Now dilute to about 40 cc. and add ammonia in slight excess. Again boil until the ammonia odor is faint and then add a marked excess of 80 per cent. acetic acid and continue the boiling for about a minute

---

* Author's modification.

longer. This last boiling is important, as it effects the neutralization or expulsion of any remaining oxidizing compounds that would cause a return of the blue color after titration. See that no copper salt remains undissolved. Cool to room temperature and add 6 cc. of a 50 per cent. solution of potassium iodide, or 3 grams of the solid. Cuprous iodide will be precipitated and iodine liberated according to the reaction,

$$2(Cu.2C_2H_3O_2)+4KI=Cu_2I_2+4(K.C_2H_3O_2)+2I.$$

The free iodine colors the mixture brown. Titrate at once with the thiosulphate solution until the brown tinge has become faint and then add sufficient starch solution to produce a marked blue coloration. Continue the titration cautiously until the last faint lilac tint is entirely removed by a single drop. In the case of an ore the presence of lead or bismuth may change the color of the final tint, but the end-point is equally sharp, especially if the last few drops of thiosulphate are allowed to fall into the center of the slowly rotating liquid, and any change from the surrounding surface noted. One cc. of the thiosulphate solution will be found to correspond to about 0.005 gram of copper, or about 1 per cent. in the case of an ore where 0.5 gram has been taken for assay. The reaction between the thiosulsulphate and the iodine is

$$2(Na_2S_2O_3)+2I=2NaI+Na_2S_4O_6.$$

Sodium iodide and tetrathionate are formed. The thiosulphate solution made from the pure crystals and distilled water is quite stable. There is usually a slight decomposition during the first 24 hours, due to the action of dissolved carbon dioxide or oxygen in the water, but after this period the solution will remain practically unchanged indefinitely if kept in a bottle of amber glass or in the dark. In my experience, actinic light appears to be the only active decomposing agent of consequence.

2. **Starch Solution.**—The following solution is the result of personal experiment: Make a cold saturated solution of commercial sodium chloride in distilled water and filter it. To 500 cc. of this solution add 100 cc. of 80 per cent. acetic acid and 3 grams of starch. Mix cold. Boil until nearly clear; about two minutes. Add a little water to replace that lost by boiling, perhaps 25 cc. A true solution of all the starch is thus obtained. No filtering or settling is required and the solution may be cooled and used at once. It keeps indefinitely and gives sharper end-points than the ordinary starch liquor.

3. **Treatment of an Ore.**—To 0.5 gram of the ore in an 8-oz. flask add 10 cc. of strong hydrochloric acid and 5 cc. of strong nitric acid. Boil until decomposition is complete, using more of the acids if necessary (and enough at the end to hold all soluble salts in solution), and then add 8 cc. of strong sulphuric acid and boil to abundant fumes, best over a free flame, but do not boil off much of the sulphuric acid.* After cooling, boil with about 30 cc. of water for a moment and then allow to stand, hot, until any anhydrous ferric sulphate has entirely dissolved; then filter through a 9-cm. filter to remove more especially any lead sulphate. Receive the filtrate in a beaker about 6-cm in diameter. Wash the filter and residue at least six times with hot water. The final volume of the filtrate will ordinarily not much exceed 75 cc. Place in the beaker a piece of aluminum prepared as follows: Cut a strip of stout sheet aluminum 2.5 cm. wide and about 14 cm. long and bend this into a triangle so it will stand on its edge in the beaker. The same aluminum may be used repeatedly as it is but little attacked

* If impure sulphur separates, it is best cleaned by allowing the strong sulphuric acid to continue just at a boil (so as not to evaporate much) for some time.

each time. Cover the beaker and heat to boiling. Allow to boil from 7 to 10 minutes, which should be sufficient to precipitate all the copper in any case, provided the volume of the solution does not much exceed 75 cc. Avoid boiling to very small bulk, as in that case some of the precipitated copper may redissolve. The aluminum should now appear clean, the precipitated copper being detached or only loosely adhering. Remove from the heat and wash down the cover and sides of the beaker with hydrogen sulphide water. This will prevent any of the finely divided copper from becoming oxidized and dissolved and will also precipitate any traces of copper that may still remain in solution. (It should be understood that there is no difficulty in precipitating *all* the copper in every case.) If the hydrogen sulphide shows there was more than a very little copper still in solution it is best to continue the boiling for a short time, diluting the liquid to about 75 cc. if necessary. This will coagulate the sulphide and render it easier to filter. Finally, decant through a 9-cm. filter and then, without delay, transfer the precipitate to the filter with the aid of a jet of hydrogen sulphide water, leaving the aluminum, as clean as possible, in the beaker. Wash filter and precipitate six times with hydrogen sulphide water, allowing to drain completely each time, but never permitting the filter to remain thus drained until finished, or the copper may oxidize and run through. Now place the original clean flask under the funnel and then open the filter carefully and spread it smoothly in the funnel. With a jet of hot water, using as little as possible, wash the precipitate into the flask. Allow 5 cc. of strong nitric acid to run from a small pipette over the aluminum in the beaker, and then pour from the beaker over the filter to dissolve any remaining copper. Without washing beaker or filter at this stage, collect the filter into small compass with a glass rod and push it gently into the throat of the funnel. Now remove the flask and replace it with the beaker. Boil the mixture in the flask to dissolve the copper and expel the red fumes, and then again place the flask under the funnel. Now pour over the filter 5 cc. or more of saturated bromine water, enough to impart a distinct color to the liquid in the flask. The bromine cleanses any residue still adhering to the filter, but its most important function is to insure the highest state of oxidation of any arsenic or antimony in the solution. Next, wash the beaker and aluminum, pouring through the filter, and then wash the filter six times with hot water. If the operations are properly conducted, the final volume of the liquid in the flask need not much exceed 40 cc. Boil the solution until the excess of bromine is entirely expelled and the volume is reduced to perhaps 25 cc. (A boiling rod (1, 14) will entirely prevent bumping.) Now add a slight excess of ammonia (usually 10 cc. of strong ammonia), boil off most of the excess, add acetic acid in excess, boil a minute, cool, dilute to about 30 cc. if necessary, add potassium iodide (see 5), and titrate as in the standardization of the thiosulphate, calculating the percentage of copper accordingly.

4. It is best to boil off as much of the free ammonia as possible, before acidifying with acetic acid, in order to avoid the formation of too much ammonium acetate, which has a retarding effect on the reactions of the subsequent titration. The point cannot be told, if much copper is present, by the absence of an ammoniacal odor, as the blue copper-ammonium salt gradually decomposes and gives off ammonia. It is usually best to add the acid when the smell of ammonia has become rather faint. Boiling too long does no harm unless a bluish deposit forms on the flask. This may be copper arsenate or hydroxide, which frequently dissolves with great difficulty in acetic acid. Always look for this deposit before adding the acetic acid, and, when observed, cautiously add enough ammonia to dissolve, it, again boiling off any great excess. A precipitate of any nature, due to overboiling, which

does not deposit on the flask, will usually do no harm, and the copper contents will be subquently taken up by the acid. A light-colored flocculent precipitate, sometimes observed in the acetic acid solution, perhaps aluminum hydroxide, is ordinarily without effect. If the cooling, before titration, causes the separation of crystals of copper acetate, they should be redissolved by warming with a little more water and the solution again cooled.

5. **Notes.** — According to the equation previously given, 0.5 gram of copper requires 2.61 grams of potassium iodide. While direct experiment shows this to be apparently true, yet when only the theoretical amount of potassium iodide is present the reaction is slow, and in fact does not appear to proceed to completion until during the titration, which is thereby unduly prolonged. It is best, therefore, to always use an excess, but as the iodide is expensive the quantity used should be governed by the amount of copper present, which can always be estimated approximately. Allow, say, 1 gram of potassium iodide for every 15 per cent. copper, when 0.5 gram of ore is taken for assay. It is convenient to prepare a solution containing 50 grams of potassium iodide in 100 cc. A 2-cc. pipette will thus deliver 1 gram of the salt. No error will be introduced in a doubtful case by adding more potassium iodide after the titration is apparently finished and resuming the operation if the blue color is thereby restored.

6. Zinc and silver do not interfere with the assay. Lead and bismuth are without effect, except that by forming colored iodides they may mask the approach of the end-point before adding starch. Arsenic and antimony, under the treatment described, have no influence. The return of the blue tinge in the titrated liquid after long standing is of no significance, but a quick return, which an additional drop or two of the thiosulphate does not permanently destroy, is usually an evidence of faulty work.

7. In such as case, or where the end-point has been accidentally passed, it is not necessary to begin the assay anew. The following procedure will enable the titration to be repeated: * Add 10 cc. of strong nitric acid and heat the mixture to boiling. Heat very cautiously at first until the iodine set free is mostly expelled, otherwise the mixture may foam over. Now manipulate the flask in a holder over a free flame and boil the solution down as rapidly as desired until only about 5 cc. are left. Dilute with 25 cc. of hot water and boil again for a short time to expel any red fumes. Now add a slight excess of ammonia and finish in the usual way.

8. **Rapid Assay by the Iodide Method.** — The following scheme has frequently proved useful where extreme accuracy was not necessary. As a rule the results are a trifle low, but the error does not usually exceed 0.10–0.15 of 1 per cent.

Treat 0.5 gram of the ore in an 8-oz. Erlenmeyer flask with 10 cc. of strong nitric acid. Heat gently until the decomposition of the copper compounds is complete, then add 5 cc. more of nitric acid and 5 grams of potassium chlorate. Continue the heating, either on a hot plate or with a thin asbestos sheet under the flask (1, 13), just to complete dryness, avoiding overheating and baking. Allow to cool, add 25 cc. of strong ammonia and 10 cc. of water. Heat gently to effect thorough disintegration of the residue and solution of the copper and then filter through a 9-cm. filter, washing with cold water. Receive the filtrate in an 8-oz. flask. Boil until the excess of ammonia is expelled and the liquid reduced to perhaps 25 cc., then acidify with acetic acid, boil a minute and finish in the usual manner.

---

* Another, and easier, method is to add, from a burette, a counted number of drops of permanganate solution, of any strength, until the color returns. Titrate, repeat the same addition of permanganate, again titrate and deduct.

9. Electrolytic Method.—To 1 gram of the ore in an 8-oz flask add 10 cc. of strong hydrochloric acid and 5 cc. of strong nitric acid. Boil gently until decomposition is complete, using more of the acids if necessary. If the evaporation of the liquid causes a separation of soluble salts, add enough hydrochloric acid at the end to again bring them into solution. Finally, add 7 cc. of strong sulphuric acid and boil, best over a free flame, to abundant fumes. After cooling, boil with about 30 cc. of water for a moment and then allow to stand, hot, until any anhydrous ferric sulphate has entirely dissolved. If silver is liable to be present add a single large drop of strong hydrochloric acid just previous to the boiling, to precipitate it as chloride. Filter, washing filter and residue at least six times with hot water. Dilute the filtrate to about 300 cc. (the dilution is unnecessary if the amount of copper is small) and pass in hydrogen sulphide until the copper is all precipitated, as shown by the clear condition of the supernatant liquid. Filter off the sulphides, washing well with hydrogen sulphide water. Now rinse the precipitate back into the beaker as completely as possible with hot water, place the beaker under the funnel and pour through the latter a hot mixture of 5 cc. of a moderately strong colorless solution of sodium sulphide (see p. 29) and 15 cc. of water. Remove the beaker, stir the contents well, and, either immediately or after warming a few minutes, according to the amount of arsenic and antimony apparently present, filter through the last filter again. A second extraction is rarely necessary. Wash out the beaker with hot water and then wash the precipitate well with hot dilute sodium sulphide solution. Reserve the filtrate, which almost invariably contains a little dissolved copper. Rinse the residue back into the beaker with hot water, using as little as possible. It is difficult to get the filter very clean and it is not necessary. Place beaker and contents once more under the funnel and pour through the latter a hot solution of about 2 grams of potassium cyanide in about 15 cc. of water. Remove the beaker and replace it with the original flask. Heat and stir the mixture in the beaker until the copper sulphide is all dissolved, adding more potassium cyanide if required, but avoiding a great excess. Sulphides of bismuth, cadmium, etc., will remain undissolved. Again filter through the last filter and wash ten times with hot water. Use small washes each time to avoid a bulky filtrate. Remove flask and filtrate to a hood, add 5–6 cc. of strong sulphuric acid (this is usually a sufficient excess), heat to boiling and boil to fumes. Finally, finish over a free flame until nearly all the excess of acid is expelled. Allow to cool with the flask inclined, to prevent the cake which may form from cracking the glass. The addition of a little nitric acid to the sulphuric acid mixture causes a more rapid solution of the copper sulphide, but this is a disadvantage as the mixture is then liable to bump badly. Dissolve the cool cake or residue in about 25 cc. of water. During these operations attention may be turned to the reserved filtrate, which usually contains an appreciable amount of copper. Acidify it with hydrochloric acid, stir well to coagulate the precipitated sulphur, filter, and wash with hydrogen sulphide water. Ignite the filter and contents in a platinum dish at a dull red heat until the carbon of the paper is entirely consumed. Be very careful to do this at as low a temperature as possible. All the arsenic and antimony will be expelled. Warm the residue with a few drops of nitric acid and add the solution to the main portion in the flask. To the latter now add 3 cc. of strong nitric acid, transfer the solution to a suitable beaker, dilute, and electrolyze as described below.

10. Electrolysis of the Copper Solution.—The solution should have a volume of about 125 cc. and contain several cubic centimeters of strong nitric acid (sp. gr. 1.42). The amount of free nitric acid necessary is not narrowly limited. It is gradually changed to ammonia by the electrolysis, and therefore, if too little be present, the solution may become alkaline. On the other hand, too much acid retards or may prevent the deposition of the copper. I usually use an excess of about 3 cc. in the above volume of liquid, but 7 or 8 cc. may be safely employed, with the advantage of preventing in a large measure the deposition of any arsenic that may have escaped extraction.

The following apparatus may be used: A cathode consisting of a plain platinum cylinder 5 cm. long and 2.5 cm. in diameter. It has a total surface of about 78.5 sq. cm. and weighs about 12.5 grams. It is supported by a strong platinum wire attached to the top. An anode consisting of a stout platinum wire rising from the center of a base made by coiling the wire around itself closely so as to form a circular disc. It weighs about 8.5 grams. A beaker suitable for the above electrodes, having a diameter of about 5 cm. and a height of about 8–9 cm.

The volume and acidity of the solution having been properly adjusted and the cathode cleaned, ignited, and weighed, the electrodes are inserted and suitably supported in the beaker and the latter is covered with a split watch-glass, which permits the wires to pass through the center, leaving an opening of only a small crack. Adjust the anode within the cathode with its base almost touching the bottom of the beaker. The bottom of the cathode should come about one-fourth of an inch above the base of the anode, and the top of the cylinder should project a little above the surface of the solution.

Now connect with the battery and electrolyze. The cathode should be connected with the zinc pole of the battery. The current density should be $ND_{100} = 0.5-1$ amp. Electrode tension, 2.2–2.5 volts. Temperature, 20°–30° C. Time required, 4 to 5 hours. For details as to the attainment of these conditions see article ELECTROLYSIS, p. 10. (In my own work, using a direct 220-volt current from a dynamo and a bank of lamps as a resistance—which reduces the tension to about 2.2 volts—I usually employ, with the above apparatus, from 0.2 to 0.4 amp., according to the amount of copper present, and allow to run over night.)

When the solution has become colorless and the copper is apparently all deposited, immerse the cathode deeper in the liquid, or better, raise the level of the latter by rinsing the cover and sides of the beaker, and allow the current to run for half an hour longer. The fresh platinum surface will show whether copper still remains in solution. Finally, at the end, lower the beaker from the electrodes, with the current still passing, and at the same time rinse off the adhering acid solution with a stream from the wash-bottle. Immediately replace the beaker with another of distilled water and then stop the current. Remove the cathode and wash off the water with alcohol. Allow to drain a moment on filter-paper and then dry at about 100° C., cool to room temperature, and weigh. The excess over the original weight of the electrode represents the copper in the weight of ore taken.

It is always a good plan to clean the electrode again with nitric acid, and after igniting and weighing it, replace it once more in the solution and electrolyze for a short time to make certain that all the copper is removed.

The above description will serve very well for those who have to make electrolytic copper assays only occasionally. By means of special arrangements of a less simple nature, including the rapid rotation of one of the electrodes, all the copper in an assay can be satisfactorily deposited very quickly, perhaps in 15 minutes. Detailed descriptions of experiments in this direction can be found in the chemical literature of the past few years. (See foot-note, p. 15.)

11. Electrolytic Methods of the Anaconda Copper Mining Co.*—Allow about 0.11 ampere for each determination. With the present arrangements at Anaconda this gives a tension of from 1.40 to 2.30 volts.

In all cases a drop of the electrolyte is brought in contact with a drop of hydrogen sulphide water to test if copper has been completely deposited.

The platinum cylinders with deposited copper are washed in water, then alcohol, dried on a steam-bath, and weighed.

If the deposited copper is dark from presence of arsenic, it is dissolved in 8 cc. of nitric acid, diluted with water and electrolyzed again, care being taken to remove from the current soon after the complete deposition of the copper.

Copper may be separated from bismuth, antimony, and arsenic by precipitation as sulphocyanate. The precipitate is washed thoroughly, ignited gently in a porcelain crucible, dissolved in 7 cc. of nitric acid, and the copper is then determined electrolytically.

In samples containing arsenic, antimony, tellurium, and selenium, such as electrolytic slimes, add 100 mgs. of iron to the nitric acid solution of the sample from which silver has been removed as chloride. Add ammonia to precipitate iron and still have the solution acid. Bring to a boil, settle, filter off the iron precipitate, redissolve it, and again precipitate as before. A third precipitation may be necessary to be certain of having all the copper in solution. Combine the filtrates, evaporate sufficiently, add nitric acid, and electrolyze. The iron precipitate holds the arsenic, antimony, tellurium, and selenium.

Converter Copper.—Dissolve 0.5 gram in 8 cc. nitric acid, 8 cc. water and 1 cc. sulphuric acid, keeping the beaker covered

* F. W. Traphagen, Western Chem. and Met., Vol. 6, 192.

during solution. When solution is complete and the fumes expelled, dilute with water and electrolyze.

In the copper determinations on mattes and converter copper, the percentage of silver, as determined by fire assay, is deducted from the combined percentages of copper and silver found by electrolysis. (291.66 oz. = 1 per cent.) The silver may be precipitated with just sufficient dilute sodium chloride solution, taking care to avoid an excess, and after filtering off the silver chloride, the copper deposited alone.

Mattes.—Moisten 1 gram of the sample with a few drops of water, add 8 cc. of nitric acid and 1 cc. of sulphuric acid. Run to dryness on steam-bath. Take up with water and 8 cc. of nitric acid. Filter and electrolyze.

Slags.—Decompose 2 grams in a platinum dish with 8 cc. nitric acid, 8 cc. hydrofluoric acid, and 1 to 2 cc. sulphuric acid. Evaporate to sulphuric acid fumes. Take up with water and 10 cc. of nitric acid and electrolyze.

Sulphide Ores.—Weigh 1 gram of the sample into a beaker ($3\frac{1}{2}$ inches high and $2\frac{1}{4}$ inches in diameter). Add 8 cc. of nitric acid and a little potassium chlorate. Evaporate to complete dryness on steam bath. Take up with water and from 6 to 10 cc. of nitric acid. Dilute sufficiently, allow to settle and then electrolyze.

Oxidized Ores.—Take 1 gram of the sample. Evaporate to dryness with 8 cc. of nitric acid. Add 10 cc. of hydrochloric acid and 2 cc. of sulphuric acid, and evaporate to sulphuric acid fumes. Take up with water and 8 cc. of nitric acid, dilute sufficiently, settle, and electrolyze.

12. Rapid Electrolytic Method.—By employing a strong current copper can be deposited very rapidly, but, in the ordinary electrolytic cell, much of the deposit is then apt to be non-coherent or detached and quite unfit for handling and weighing. This trouble has been overcome by the use of arrangements which rotate one of the electrodes or the electrolyte. These devices are complicated and costly. Quite recently it has been found that almost equal rapidity may be attained by the use of a specially prepared gauze cathode.* This electrode is made of 52-mesh platinum wire gauze. It is about 1 inch in diameter and 1½ inches long, of cylindrical shape and corrugated and sand-blasted. With 75 cc. of solution and a current of from 8 to 10 amperes and 3 to 4 volts, the copper is usually completely deposited in from 10 to 15 minutes. The character of the deposit on the prepared electrode is all that could be desired.

Ores may be treated as described in 9 until the copper, free from interfering elements, is finally obtained in solution as sulphate. Neutralize the excess of sulphuric acid with ammonia, then add an excess of 3 cc. of strong nitric acid, transfer to a tall electrolysis beaker, dilute to 75 cc. and electrolyze with a current of the above strength. The end of the operation may be determined by testing a drop of the electrolyte with a drop of hydrogen sulphide water.

In the absence of interfering elements the process may be greatly simplified, even omitting filtration when the insoluble residue is slight or settles well. The final conditions, however, as described above, must always be the same, and the solution must be free from chlorides.

Arsenic, if present only in small amount, will not deposit with the copper. Therefore, if the deposited copper is darkened with arsenic, it usually suffices to redissolve it in nitric acid and deposit it once more.

13. Cyanide Method.—In this method the copper is obtained in a blue ammoniacal solution and its amount is estimated from

---

* R. C. Benner, Jour. Ind. and Eng. Chem., May, 1910. Benner's electrodes may be obtained of the Denver Fire-Clay Co.

the quantity of standard solution of potassium cyanide required to discharge the blue color. The results of the cyanide titration are exact if certain conditions are always maintained. It is found that for the same amount of copper:

1. A concentrated solution requires more cyanide for decoloration than a dilute solution.

2. A hot solution requires less cyanide than a cold one.

3. In any case when, from a rapid addition of cyanide, the color has become rather faint, it may, by simple standing, continue to fade, and perhaps entirely disappear.

4. If the amount of cyanide added is insufficient to effect complete discharge of color, even after allowing the copper solution to stand for several minutes, the titration may then be finished without alteration of the final result.

From the foregoing facts it is evidently necessary, in order to obtain correct results, that the titrations for unknown amounts of copper should be made under conditions that do not differ materially in the following particulars from those governing the standardization of the cyanide solution:

1. Temperature.

2. Rapidity of the final additions of cyanide.

3. Final volume of solution.

Besides the physical conditions just enumerated, there are chemical conditions that effect the result, such as presence of a large amount of chlorides, a large excess of ammonia, etc. Such abnormal conditions require no special consideration, since they are all easily avoided by following the method to be described.

14. Standardization of the Cyanide Solution.—Dissolve pure potassium cyanide in distilled water in the proportion of 21 grams to the liter. Weigh accurately about 0.2 gram of pure copper foil and dissolve it in an 8-oz. flask in 5 cc. of strong nitric acid (sp. gr.

1.42). Dilute with 25 cc. of water and add 5 cc. of a saturated solution of bromine in cold water. Boil the mixture until the bromine is apparently expelled. Now add 50 cc. of cold water and 10 cc. of strong ammonia (sp. gr. 0.90). Cool to the ordinary temperature by placing under a tap or in cold water. Titrate with the cyanide solution in a slow, cautious manner, and as the end-point is approached, as shown by the partial fading of the blue color, add distilled water so as to bring the volume of the solution to approximately 150 cc. Finish the titration by careful and regular additions of cyanide, finally decreasing to a drop at a time and agitating the flask with a rotary movement after each addition, until the blue tint can no longer be detected by holding the flask against a light-colored background.* It is, of course, very essential that there should be no haste and no prolonged delay in these final additions of cyanide. Simply adopt a regular, natural manner, that can easily be repeated in all subsequent titrations.

From the amount of cyanide solution required for the weight of copper taken, calculate the value of 1 cc. in copper.

Keep the standard solution in a cool place not exposed to direct sunlight. Under these circumstances it holds its strength fairly well, but still it gets weaker from the decomposition of the cyanide and should be restandardized weekly.

15. **Treatment of Ores, etc.**—Treat 1 gram, or 0.5 gram if the material seems to contain 40 per cent. or over, in an 8-oz. flask with 10 cc. of strong nitric acid. Boil gently until decomposition appears to be complete and then add about 7 cc. of strong sulphuric acid and heat the mixture over a free flame until all the nitric acid is expelled and the residuary sulphuric acid is boiling freely and evolving copious fumes. Remove from the

* Some operators prefer a porcelain casserole and stirring-rod to a flask.

flame and allow to cool. Ores that are not decomposed by this treatment must be attacked in some special manner for which no general directions can be given. Sometimes the addition of hydrochloric acid is all that is necessary. It is advisable in any case not to add the sulphuric acid until the ore appears to be well decomposed.

To the residue in the flask add 20 cc. of cold water and heat the mixture to boiling. If the ore is liable to contain an appreciable amount of silver, add a single drop of strong hydrochloric acid and agitate the liquid so as to collect the silver chloride in clots. One per cent. of silver, or 292 ozs. per ton of 2000 lbs., will increase the apparent copper contents about 0.29 per cent. Allow to stand, hot, until any anhydrous ferric sulphate is entirely dissolved and then filter, washing flask and filter with hot water. Return the filtrate to the original flask. The volume of the solution should not much exceed 60 cc. at this point. Now place in the flask 3 pieces of stout sheet aluminum, each about $1\frac{1}{2}$ inches long by $\frac{5}{8}$ inch wide, and heat the mixture to boiling. Boil for perhaps 5 or 10 minutes, according to the volume of the liquid and the appearance of the aluminum. When the copper is all precipitated the aluminum will usually appear bright and clean, or it will become clean by agitating the flask so as to loosen the adhering copper. Remove from the lamp, add about 15 cc. of strong hydrogen sulphide water, which will insure the complete precipitation of the copper, allow to settle a moment, and then decant through a 9-cm. filter, retaining in the flask the aluminum and as much of the copper as possible. Wash the precipitated copper 2 or 3 times by decantation with weak hydrogen sulphide water, using about 25 cc. each time and pouring through the filter. Drain the flask as completely as possible the last time. Now place the flask under the funnel and pour through the latter 10 cc. of a warm mixture of equal volumes of strong nitric acid and water. Do not wash the filter at this point, but remove the flask and replace it with a beaker. Shake the acid about in the flask gently, so as to dissolve all the copper, warming slightly if necessary, but avoid heating more than is required to just dissolve the copper, or the aluminum may be attacked When solution is complete, pour the entire contents of the flask into the beaker that was placed under the funnel, washing only the lip of the flask, and then pour the solution back into the flask again, retaining the aluminum in the beaker. Wash the aluminum thoroughly and then replace the flask under the funnel. Now pour into the filter 5 cc. of a cold saturated solution of bromine in water, and when it has run through wash the filter with hot water. The bromine is to cleanse any dark sulphur left from the copper sulphide on the filter. In the above operations avoid increasing the bulk of the solution more than necessary. Boil the solution in the flask until the bromine is expelled, then cool somewhat and add 10 cc. of strong ammonia (sp. gr. 0.90) and then continue the cooling to room temperature. Titrate the cool solution with the standard cyanide solution cautiously until the blue color is discharged to a considerable extent and it is evident that the end-point is not far off.

The liquid is now frequently more or less cloudy. When this is the case it should, for accurate work, be filtered. If the titration has been carried too far before filtration, the faint-blue tinge is liable to fade completely away, thus spoiling the assay. On the other hand, if filtered early in the titration, a second milkiness may develop later. Filter the partially titrated solution through a 12.5-cm. filter. One washing will usually suffice. Finish the titration very carefully on the clear, pale-blue solution, precisely as in the standardization previously described. Toward the end dilute if necessary, so as to obtain a final volume of about 150 cc.

The number of cubic centimeters of cyanide solution required, multiplied by the copper value of 1 cc., will give the weight of copper contained in the amount of ore taken, from which the percentage is readily calculated.

None of the ordinary constituents of ores interfere with the method as described.

16. Permanganate Method.* — In this method the copper is precipitated as cuprous thiocyanate. This is subsequently decomposed with sodium hydroxide with the formation of sodium thiocyanate, and the solution of the latter, after acidifying with sulphuric acid, is titrated with standard potassium permanganate solution. The decomposition and titration are in accordance with the following reactions:

$$CuCNS + NaOH = CuOH + NaCNS,$$

$$10HCNS + 12KMnO_4 + 8H_2SO_4 = 6K_2SO_4 + 12MnSO_4 + 10HCN + 8H_2O.$$

The permanganate solution is made of such a strength that 1 cc. = 10 mgm. Fe. Theoretically, the factor for copper from the iron value is 0.1897, but, owing to the slight solubility of the cuprous thiocyanate, the actual factor for small percentages of copper under the above conditions is 0.200.

17. Treatment for Low-grade Material.—Weigh 2 grams into a 150-cc. beaker, add 8 cc. of strong nitric acid and evaporate to dryness. In most cases simple dryness is sufficient, but if gelatinous silica is present, the drying should be more complete. To the residue add 2 cc. of hydrochloric acid, heat to boiling, add 20 cc. of hot water and filter. Heat the filtrate and add 10 cc. of a 10 per cent. sodium sulphite solution, and when the liquid is colorless add 5 cc. of 10 per cent. potassium thio-

* G. A. Guess, private communication, and Jour. Am. Chem. Soc., XXIV.

cyanate solution and heat to boiling to coagulate the precipitate. Filter through a double 11-cm. filter and wash thoroughly with hot water to remove all soluble thiocyanates. Now place a clean beaker or flask under the funnel and thoroughly wash the filter and precipitate once with boiling 8 per cent. sodium hydroxide solution, contained in a wash-bottle, and then wash thoroughly with hot water. Finally, make the alkaline filtrate decidedly acid with dilute sulphuric acid and titrate to the usual pink tinge with standard potassium permanganate.

The only point to watch carefully is the acidity at the time of adding the sodium sulphite and thiocyanate. When only a few milligrams of copper are present the amount of free acid should not exceed about 2 per cent., as otherwise the precipitation may fail to take place. In such a case the excess of acid may be neutralized by the further addition of sodium sulphite after adding the thiocyanate. With ordinary care, however, the trouble will not occur. No elements interfere.

18. Treatment for Ores.—Weigh 0.5 gram of the ore into a 300-cc. beaker, add 5 cc. of strong nitric acid, heat until decomposition is complete and then evaporate until the beaker is almost dry, finally adding, while the beaker is still over the heat, 3 cc. of strong hydrochloric acid, followed by 75 cc. of hot water. If silver is present in appreciable amount, it would cause an error in the result unless removed by filtration at this point, but if absent or negligible, as is usually the case, the filtration may be omitted. To the hot solution, or filtrate, add 5 cc. of a 10 per cent. solution of sodium sulphite, and then, after decolorization of the liquid, 5 cc. of a 10 per cent. solution of potassium thiocyanate. Heat to boiling to effect coagulation of the precipitate and then filter and wash thoroughly with hot water, using a double 11-cm. filter.

Decompose the precipitate with sodium hydroxide solution and finish the assay as described in 17.

Cc.	% Cu.	Cc.	% Cu.	Cc.	% Cu.	Cc.	% Cu.	Cc.	% Cu.
0.1	0.04	5.1	2.09	10.1	4.13	15.1	6.16	20.1	8.18
0.2	0.08	5.2	2.12	10.2	4.17	15.2	6.20	20.2	8.22
0.3	0.12	5.3	2.18	10.3	4.21	15.3	6.24	20.3	8.26
0.4	0.16	5.4	2.23	10.4	4.25	15.4	6.28	20.4	8.30
0.5	0.20	5.5	2.27	10.5	4.29	15.5	6.32	20.5	8.34
0.6	0.24	5.6	2.31	10.6	4.33	15.6	6.38	20.6	8.38
0.7	0.28	5.7	2.35	10.7	4.37	15.7	6.42	20.7	8.42
0.8	0.32	5.8	2.39	10.8	4.41	15.8	6.45	20.8	8.46
0.9	0.37	5.9	2.42	10.9	4.45	15.9	6.49	20.9	8.50
1.0	0.41	6.0	2.46	11.0	4.49	16.0	6.53	21.0	8.54
1.1	0.45	6.1	2.50	11.1	4.53	16.1	6.56	21.1	8.58
1.2	0.49	6.2	2.54	11.2	4.55	16.2	6.60	21.2	8.62
1.3	0.53	6.3	2.58	11.3	4.62	16.3	6.64	21.3	8.66
1.4	0.57	6.4	2.63	11.4	4.66	16.4	6.68	21.4	8.70
1.5	0.61	6.5	2.67	11.5	4.70	16.5	6.72	21.5	8.74
1.6	0.64	6.6	2.71	11.6	4.74	16.6	6.76	21.6	8.78
1.7	0.68	6.7	2.75	11.7	4.78	16.7	6.80	21.7	8.82
1.8	0.72	6.8	2.79	11.8	4.82	16.8	6.84	21.8	8.86
1.9	0.77	6.9	2.83	11.9	4.86	16.9	6.88	21.9	8.90
2.0	0.82	7.0	2.87	12.0	4.90	17.0	6.93	22.0	8.96
2.1	0.86	7.1	2.91	12.1	4.94	17.1	6.97	22.1	9.00
2.2	0.90	7.2	2.95	12.2	4.98	17.2	7.01	22.2	9.04
2.3	0.93	7.3	2.99	12.3	5.02	17.3	7.05	22.3	9.08
2.4	0.97	7.4	3.04	12.4	5.06	17.4	7.09	22.4	9.12
2.5	1.00	7.5	3.08	12.5	5.10	17.5	7.13	22.5	9.16
2.6	1.04	7.6	3.13	12.6	5.14	17.6	7.17	22.6	9.20
2.7	1.08	7.7	3.17	12.7	5.18	17.7	7.21	22.7	9.24
2.8	1.14	7.8	3.20	12.8	5.22	17.8	7.25	22.8	9.28
2.9	1.18	7.9	3.24	12.9	5.26	17.9	7.29	22.9	9.32
3.0	1.23	8.0	3.27	13.0	5.31	18.0	7.33	23.0	9.36
3.1	1.27	8.1	3.31	13.1	5.35	18.1	7.37	23.1	9.40
3.2	1.28	8.2	3.35	13.2	5.39	18.2	7.41	23.2	9.44
3.3	1.36	8.3	3.39	13.3	5.43	18.3	7.45	23.3	9.48
3.4	1.40	8.4	3.43	13.4	5.47	18.4	7.49	23.4	9.52
3.5	1.45	8.5	3.47	13.5	5.51	18.5	7.53	23.5	9.56
3.6	1.49	8.6	3.51	13.6	5.55	18.6	7.57	23.6	9.60
3.7	1.53	8.7	3.55	13.7	5.59	18.7	7.61	23.7	9.64
3.8	1.57	8.8	3.60	13.8	5.63	18.8	7.65	23.8	9.68
3.9	1.60	8.9	3.64	13.9	5.67	18.9	7.69	23.9	9.71
4.0	1.64	9.0	3.68	14.0	5.72	19.0	7.74	24.0	9.74
4.1	1.68	9.1	3.72	14.1	5.76	19.1	7.78	24.1	9.78
4.2	1.72	9.2	3.76	14.2	5.80	19.2	7.82	24.2	9.82
4.3	1.77	9.3	3.80	14.3	5.84	19.3	7.86	24.3	9.86
4.4	1.81	9.4	3.84	14.4	5.88	19.4	7.90	24.4	9.90
4.5	1.85	9.5	3.88	14.5	5.92	19.5	7.94	24.5	9.94
4.6	1.89	9.6	3.92	14.6	5.96	19.6	7.98	24.6	9.98
4.7	1.93	9.7	3.96	14.7	6.00	19.7	8.02	24.7	10.02
4.8	1.97	9.8	4.01	14.8	6.04	19.8	8.06	24.8	10.06
4.9	2.01	9.9	4.05	14.9	6.08	19.9	8.10	24.9	10.10
5.0	2.05	10.0	4.09	15.0	6.12	20.0	8.14	25.0	10.14

Cc	% Cu.	Cc.	% Cu.	Cc.	% Cu	Cc.	% Cu	Cc.	% Cu
25.1	10.18	30.1	12.06	35 1	14.06	40 1	16.04	45.1	18.04
25.2	10.22	30.2	12.12	35 2	14.10	40 2	16.08	45 2	18.08
25 3	10 26	30.3	12.14	35 3	14.14	40 3	16.12	45.3	18.12
25.4	10.30	30.4	12.18	35 4	14.18	40.4	16.16	45.4	18.16
25.5	10 34	30.5	12 22	35 5	14.22	40 5	16.20	45 5	18.20
25.6	10 38	30.6	12.26	35 6	14.26	40 6	16.24	45 6	18.24
25.7	10 42	30.7	12.30	35.7	14 30	40 7	16.28	45.7	18.28
25.8	10 46	30.8	12 34	35.8	14 34	40.8	16.32	45 8	18.32
25.9	10 50	30.9	12 38	35.9	14 38	40 9	16.36	45 9	18.36
26.0	10 53	31.0	12.42	36 0	14.42	41.0	16.40	46 0	18.40
26.1	10 56	31.1	12.46	36.1	14.46	41 1	16.44	46.1	18.44
26.2	10.60	31.2	12 50	36.2	14.50	41.2	16.48	46 2	18 48
26 3	10 64	31.3	12.54	36.3	14 54	41 3	16.52	46 3	18.52
26.4	10 68	31.4	12 58	36.4	14 58	41 4	16.56	46 4	18.56
26 5	10.72	31.5	12 62	36 5	14.62	41 5	16.60	46 5	18.60
26.6	10.76	31.6	12.66	36.6	14 66	41 6	16.64	46 6	18.64
26 7	10.80	31.7	12.70	36.7	14.70	41 7	16 68	46 7	18.68
26 8	10 84	31.8	12 74	36 8	14 74	41 8	16.72	46 8	18.72
26 9	10.88	31.9	12 78	36.9	14 78	41 9	16 76	46 9	18.76
27.0	10 92	32.0	12 82	37.0	14.82	42.0	16 80	47.0	18.80
27.1	10 95	32.1	12 86	37.1	14.87	42 1	16.84	47.1	18.84
27.2	10 98	32.2	12 90	37.2	14 91	42 2	16.88	47.2	18.88
27 3	11.02	32.3	12.95	37.3	14.95	42.3	16.92	47.3	18.92
27.4	11 06	32 4	12 99	37.4	14.99	42.4	16 96	47.4	18.96
27.5	11 10	32.5	13.04	37.5	15.03	42.5	16.99	47.5	19.00
27 6	11 14	32 6	13 08	37.6	15.07	42 6	17.04	47.6	19.04
27.7	11.18	32.7	13 12	37.7	15.11	42 7	17.08	47.7	19.08
27.8	11 22	32.8	13 16	37.8	15.13	42.8	17.12	47.8	19.12
27.9	11 26	32.9	13 20	37.9	15.16	42 9	17.16	47.9	19.16
28 0	11.30	33 0	13 23	38.0	15.20	43 0	17.20	48.0	19.20
28.1	11 33	33.1	13 27	38.1	15.24	43 1	17.24	48.1	19.24
28.2	11.37	33 2	13 31	38.2	15.28	43.2	17.28	48.4	19.28
28.3	11.41	33.3	13.35	38.3	15.32	43 3	17.32	48.3	19.32
28.4	11 44	33.4	13.39	38.4	15.36	43.4	17.36	48.4	19.36
28 5	11 48	33 5	13.43	38.5	15.40	43 5	17.40	48.5	19.40
28 6	11 51	33.6	13 47	38.6	15.44	43.6	17.44	48.6	19.44
28 7	11 55	33.7	13.51	38.7	15.48	43 7	17.48	48 7	19.48
28.8	11.59	33 8	13 55	38.8	15.52	43 8	17 52	48.8	19.52
28.9	11.60	33 9	13 59	38.9	15.56	43.9	17.56	48 9	19.56
29.0	11.65	34.0	13.63	39.0	15.60	44 0	17.60	49.0	19.60
29.1	11.69	34.1	13.67	39.1	15.64	44.1	17.64	49.1	19.64
29.2	11.73	34 2	13.71	39.2	15.68	44.2	17.68	49.2	19.68
29 3	11.77	34.3	13 75	39.3	15.72	44.3	17.72	49.3	19.72
29.4	11.81	34.4	13 79	39.4	15.76	44.4	17.76	49.4	19.76
29.5	11.85	34.5	13 83	39.5	15 80	44.5	17.80	49.5	19.80
29.6	11 88	34.6	13.87	39.6	15.84	44.6	17.84	49.6	19.84
29 7	11.92	34.7	13 91	39.7	15.88	44.7	17.88	49.7	19.88
29.8	11.94	34.8	13 95	39.8	15.92	44.8	17.92	49.8	19.92
29.9	11 98	34.9	13 99	39.9	15.96	44.9	17.96	49.9	19.96
30.0	12 02	35 0	14 03	40.0	16.00	45.0	18.00	50 0	20.00

Owing to the slight error introduced by the solubility of cuprous thiocyanate, Guess titrates with a permanganate solution of the exact strength 1 cc. = 0.010 gram Fe, and makes use of the accompanying table, which has been carefully prepared after long comparison with the electrolytic method on the same samples.

With low-grade material, where 2 grams are taken for assay, the error may be neglected and the copper factor deduced as 0.2 of the iron value.

19. Guess' Electrolytic Method for Ores, etc.—Mr. G. A. Guess observed, in the course of some electrolytic copper work, that a "nitro" preparation, formed by the action of strong nitric acid on a certain petroleum product, when added to the usual nitric acid electrolyte, permitted the employment of a strong current, with a corresponding shortening of the time of deposition, without injuring the reguline nature of the deposit. Thus the usual 8 to 12 hours was reduced to 3. At the same time it was found that even large amounts of arsenic and antimony did not interfere and the deposited copper remained bright and uncontaminated.

The following are the details of the method:

*The Nitro Compound.**—Heat very gently a mixture of about 10 grams of vaseline and 100 cc. of strong nitric acid. When all action has ceased, dilute to about 300 cc. and allow to stand for 24 hours. After filtering the straw-colored liquid it is ready for use.

The Electrodes.—These are of the Guess-Haultain design and are made of 0.001-inch platinum foil. The cathode is 12.5 cm. long and is divided into a blade 4 cm. wide and 6.25 cm. long, and a central tongue 0.7 cm. wide and 6.25 long, the immersion area being 50 sq. cm. and weight 1.5 grams. The blade is first

* Modification of Messrs. Cavers and Chadwick. Eng. and Min. Jour., Vol. 89, 954.

sand-blasted and then corrugated lengthwise, in order to impart the necessary rigidity. The sand-blasting is to permit a firmer adhesion of the deposited copper, which is otherwise liable to fly off during the final drying.

The anodes are 12.5 cm. long and 0.5 cm. wide with a median corrugation. Three electrodes are used in each cell; one cathode in the middle and one anode on each side of the cathode. These electrodes are connected to slotted aluminum terminals, in which they are held by contact pressure. The terminals are ⅝-inch rods, projecting 2 inches horizontally in front of the wall of the cabinet; at the back, the middle electrode (cathode) is connected with one pole of the current, and the two outer ones (anodes) with the other pole.

The Procedure.—Weigh the ore into a tall narrow beaker of about 200 cc. capacity, suitable for the electrolysis. Digest with 7 cc. of nitric acid and boil until the red fumes are expelled. Add about ½ cc. of the prepared nitro compound, nearly fill the beaker with water and allow to stand and settle for a moment. Insert the electrodes and electrolyse with a current of 1.5 amperes for 3 hours.

There should be no evolution of gas whatever at the cathode during the electrolysis. Gas is frequently evolved when the current is first turned on, but if turned off for a second and then on again, the bubbles should cease. Gas may again appear at the end of three hours, when the assay is finished.

The cathode with the deposited copper is finally removed, dried, and weighed in the usual manner (10).

20. The Colorimetric Determination of Copper.*—The colorimetric method is applicable only in the case of products containing small amounts of copper, such as slags and tailings. It is

* From papers by F. W. Traphagen, Western Chem. and Met., VI, 148, and Arthur Austin, Western Chem. and Met., VI, 309.

based upon the depths of color produced by cupro-ammonium nitrate, and, with proper precautions, it is a delicate and accurate means of determining small percentages of copper. To secure accuracy it is necessary to have the conditions under which the comparison or standard solution was obtained, and that of the sample under examination, as nearly identical as possible. Hence, instead or preparing the standards with known quantities of pure copper only, the various other impurities accompanying the copper in the sample are simulated, by using tailings, blast or reverberatory slags, of known copper content in making up the standard comparison solutions. The following methods are modifications of the method of Thorn Smith. They were developed and are used in the laboratory of the Anaconda Copper Mining Company.

Preparation of Standards.

Blast Slags.—Take 3 grams of sample on which the copper has been determined electrolytically, cover with water, add 10 cc. of nitric acid and 1 cc. of hydrochloric acid and heat a few minutes on the steam-bath. Dilute the mixture, after heating, with 100 cc. of water, add dilute ammonia in slight excess and filter into a colorimetric bottle. Wash until the filtrate fills the bottle to the mark. If the electrolytic copper on this sample was 0.20 per cent., this standard is called "B–2." To prepare "B–3" add 0.003 gram of copper to another 3 grams. "B–4" is prepared by the addition of 0.006 gram of copper, and "B–5" by adding 0.009 gram of copper, the copper being always added before the ammonia, and the samples treated as in the preparation of "B–2." If samples low enough in copper to prepare the lowest standards are not at hand, all the copper may be removed electrolytically and then sufficient of a standard copper solution added for the required standard.

Tailings.—Heat 1 gram of the sample on the steam-bath with 5 cc. of nitric acid and a pinch of potassium chlorate. Dilute, filter and wash as for slags. If the electrolytic copper on this sample was 0.50 per cent., the standard is called "T–5." To prepare "T–6" add 0.001 gram of copper to another gram sample; 0.002 gram for "T–7"; 0.003 gram for "T–8;" 0.005 gram for "T–10."

Reverberatory Slags.—Take 2 grams of sample, add 10 cc. of hydrochloric acid and 2 cc. of nitric acid. Heat, dilute, add ammonia, filter and wash as for blast slags. If the electrolytic copper on this sample was 0.30 per cent., this standard is called "R–3." For "R–4" add 0.002 gram of copper to another 2 grams of sample. Add 0.004 gram for "R–5," 0.006 for "R–6," and so on. Treat each as in the preparation of "R–3."

A set of standards being prepared as above, another set of bottles is arranged, each bottle being filled almost to the mark with water and 10 cc. of ammonia. A standard solution of copper, containing 0.001 gram of copper to the cubic centimeter, is run into each from a burrette until the color produced exactly matches a corresponding standard prepared as above. The burette reading is carefully noted in each case.

The following results were thus obtained:

B–2	required	4.4 cc.	R–2	required	3.2 cc.	T–3	required	2.8 cc.
B–3	"	6.7 cc.	R–3	"	4.6 cc.	T–4	"	3.8 cc.
B–4	"	9.1 cc.	R–4	"	6.1 cc.	T–5	"	4.7 cc.
B–5	"	11.1 cc.	R–5	"	7.9 cc.	T–6	"	5.7 cc.
			R–6	"	9.7 cc.	T–7	"	6.6 cc.
			R–7	"	11.4 cc.	T–8	"	7.6 cc.
			R–8	"	13.2 cc.	T–9	"	8.5 cc.
						T–10	"	9.5 cc.

By the use of this table standards may be rapidly prepared.

Determination of Copper.

Treat blast slags, reverberatory slags and tailings exactly as described in "Preparation of Standards" and match the filtrates with the standards.

Details of Operation.—Except in the case of converter slag (of which 0.5 gram is taken) all portions are 1, 2, or 3 grams.

Tailings are weighed into No. 1 beakers, 100 cc. capacity, and slags into No. 2 beakers, 200 cc. capacity.

Tailings are treated with nitric acid and a pinch of potassium chlorate, set upon the steam-plate and left for half an hour. The addition of nitric acid should be regulated so as to leave about 5 cc. after treatment on the hot plate. After the hot plate treatment about 25 to 30 cc. of cold water are added to the beakers and they are then placed in front of the color bottles which have been previously arranged. These color bottles are of square shape, to allow of easy and safe handling, and are made of clear white glass. They are 7½ inches high, with a body 2½ inches in diameter and a short neck about 1¾ inches in diameter. 1:1 ammonia is added to the beakers in such quantity as to give about the same excess as is used in the standards. Ribbed funnels 4¼ inches in diameter are set in the mouths of the bottles, and provided with filters 18½ cm. in diameter. The contents of the beakers are now poured "clean" into the filters. The beakers are washed once with warm water, which is then poured through the filters, and, after draining, the precipitate in each filter is given one good wash with warm water. The funnels are now removed and the bottles are filled up to the standard marks with water. The larger bulk of solution before filtering and the greater amount of wash water required cause the mark on the slag bottles to be higher than that on the tailings. On the described bottle the tailings mark will come about half way up, and the slag mark will be near the top. The tests, when ready, are matched against the standard colors arranged in a color stand. The latter is made of wood, painted white, and faces a window. Each standard bottle is in a separate compartment, so as not to be affected by the color of an adjacent standard. In making comparisons the standard and the unknown should be at equal distances from the partitions, as the latter have a slight effect on the color.

Results are read and reported to the nearest 0.05 per cent. by reading between the standards, which are graduated to 0.1 per cent.

Reverberatory and blast slags require different treatment. Blast slags decompose easily; reverberatory slags are best decomposed by first adding hydrochloric acid, and then, after digestion for a few minutes on the steam-plate, making an addition of nitric acid. A little water added before the acid, so as to make a paste, will prevent undecomposed slag from sticking to the beaker bottom.

Reverberatory slags, and also converter slags, should be allowed about three-quarters of an hour on the steam-plate, and, when taken up with water, they require a good stirring with a rubber "policeman." The amount of iron and silica is large, and the dilute ammonia should be added while stirring, so as best to determine the required excess.

Sometimes the color produced in a test has a greenish tinge, and the test is then apt to be read too high. Heath states that this may be due to the presences of organic matter, which impairs the complete precipitation of ferric hydroxide by ammonia. If the green tint is too pronounced the sample should be ignited.

Standards of low copper content seem to preserve their color best with a small excess of ammonia, 5 cc. Those of higher copper content require 10 to 20 cc. excess. The sample to be matched should be as near like the standard in all respects as possible.

CHAPTER XIV.

FLUORINE.

(See Appendix.)

1. **Kneeland's Method for the Determination of Fluorine in Ores and Slags.***— Fuse 0.5 or 1 gram of the material (according to the supposed amount of fluorine present) in a porcelain crucible with ten times its weight of a mixture of equal parts of sodium carbonate and potassium carbonate until the whole mass is in quiet fusion. Raise the heat to bright red and pour into an iron mold,† saving the crucible. Cool, break up the crucible into small pieces, and transfer along with the fused mass to a 6-inch agate-ware casserole (agate-ware is preferable to porcelain as diminishing the liability to subsequent "bumping"). Add 200 cc. of distilled water and digest for one hour at a temperature near the boiling-point, breaking up the fused lump with a thick glass rod. If, at the end of this time, any undecomposed lumps are noticed, remove them with the pincers and grind them in an agate mortar and wash back into the casserole with hot water.

Now boil for 10 minutes and filter through a loose filter ‡ into a beaker of about 1 liter capacity. Wash first with hot water,

* E. Kneeland, Eng. and Min. Jour., LXXX, 1212.

† Instead of an iron mold, a clean metallic dish floating on water may be used. (A. H. L.)

‡ For this filtration I would suggest a thick wad of absorbent cotton placed in a funnel and wetted. (A. H. L.)

then with a hot solution of ammonium carbonate. The residue is now discarded. Add to the filtrate 10 grams of ammonium carbonate and boil 5 minutes, and afterwards allow to stand in the cold for two hours. Filter through a loose filter (see last footnote) into a 6-inch agate-ware casserole (decanting as much as possible of the fluid) and wash with cold water once or twice. In order to eliminate the final traces of silica, add 20 cc. of an emulsion of zinc oxide in ammonia (cf. XXIV, 16), and boil, with the casserole uncovered, until the odor of ammonia is no longer detected.

Filter into a No. 5 beaker and wash with hot water. To the filtrate add a solution of calcium chloride, while stirring with a rubber-tipped glass rod, until no more precipitate is formed. Allow the precipitate to subside, and filter, washing with hot water. Test the filtrate for carbonates and fluorine with a few drops of the calcium chloride solution. Now transfer the precipitate along with the filter-paper to a platinum dish of suitable size. Dry first, then ignite at a red heat for 20 minutes. Cool, and disintegrate the mass with hot water. Add acetic acid until the solution is clear and evaporate to dryness, being careful not to scorch. Now moisten again with acetic acid and then evaporate until the odor of acetic acid is no longer perceptible.

Wash the mass into a No. 3 beaker with hot water, add hot water until the calcium acetate is all dissolved, then about 150 cc. more and stir. Digest for a few minutes at a gentle heat and filter, washing first with hot water, then with hot ammonium chloride solution, and again with hot water. Transfer the precipitate and filter-paper to a platinum dish, dry, and ignite. Cool, moisten with cold water, add 6 cc. of strong sulphuric acid, and heat for a few minutes. Cool, dilute, and transfer the contents of the dish to a No. 2 beaker. Add 5 grams of ammonium chloride, boil for a few minutes, cool, and add an excess of strong ammonia water. Then add 2 or 3 cc. of strong hydrogen peroxide solution, boil and filter. Precipitate the calcium from the filtrate with ammonium oxalate and determine the calcium as CaO in the usual manner by titration with permanganate (x, 4). Calculate the CaO to CaF_2, from which the amount of fluorine can readily be calculated.

To calculate CaO to CaF_2 multiply the percentage of CaO found by 1.392. Multiply the percentage of CaF_2 found by 0.4865 to obtain the percentage of fluorine.

(It is simpler to use the permanganate factor for fluorine. This may be obtained by multiplying the factor for CaO by 0.6770.—A. H. L.)

2. **Penfield's Volumetric Method for Fluorine in Fluorspar.***—*Solutions and Reagents.*—One-fifth normal Sodium Hydroxide solution, prepared and standardized in the usual manner.

1 cc. of this solution = 0.0234 gram CaF_2.
" " " " = 0.0114 gram F.

Alcoholic Potassium Chloride solution, prepared by dissolving 30 grams of potassium chloride in 100 cc. of water and adding 100 cc. of alcohol. 20 cc. of the solution are used for each determination.

Lacmoid Indicator, prepared by dissolving 0.2 gram of lacmoid (resorcin blue) in 100 cc. of alcohol. This indicator is red with acids and blue with alkalies.

Powdered Silica. Ignited powdered quartz or precipitated

* This description of Penfield's method is from The Chemical Engineer, Vol. III, p. 65.

silica will either of them answer the purpose, provided they are free from fluorine.

Sulphuric Acid. This must be concentrated. It is best to heat it in a well-annealed flask until it fumes strongly and then cool, stoppering the flask with a perforated cork carrying a 6-inch piece of capillary tube, until the acid is cold, and then with a solid rubber stopper.

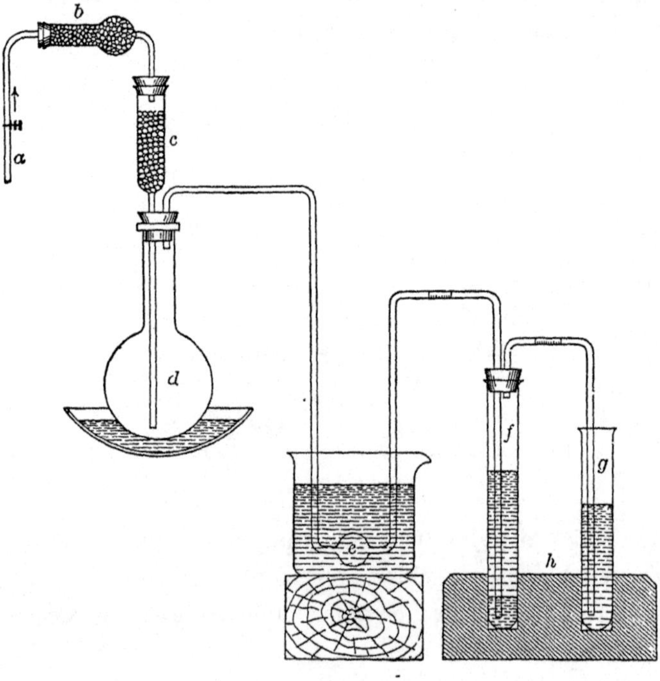

FIG. 16.

The apparatus is shown in the diagram and consists of a well-annealed 250-cc. flask *d* closed by a 2-hole rubber stopper. Through one of these holes passes a funnel, *c*, which can be easily made by joining a piece of glass tubing on to a calcium chloride tube to form a long enough stem to reach nearly to the bottom of the flask. The funnel is filled with dry glass beads, and is closed by a stopper, in the single hole of which is inserted a calcium chloride tube filled with dry granular soda-lime.

The rubber tube *a* connects the apparatus with the source of a current of air, which later can be shut off by the Hoffmann clamp as shown. In the second hole of the flask-stopper a condenser-tube *e*,* bent as shown, is inserted. This is made by blowing a bulb in a piece of glass tubing (6 mm. in diameter), and is kept cool by immersion in a beaker of cold water. The test-tube *f* is connected to *e* by a rubber joint as shown, and is provided with inlet and outlet tubes, the latter being connected with the inlet tube of another test-tube, *g*. The test-tube *f* is 8 inches high, while *g* is only 6 inches. Both tubes are held upright by a small block of wood, *h*, bored with holes to fit the tubes.

3. *Determination.*—Have the flask and condenser-tube *e* and the entrance-tube to the test-tube *f* thoroughly dry. Pour about ½ inch of mercury into the test-tube *f*, and on top of this 25 cc. of the potassium chloride solution. Half fill the test-tube *g* also with this solution.

Now weigh into an agate mortar 0.2 to 0.5 gram of the finely powdered sample, place the mortar on a piece of black glazed paper and carefully mix, by gentle rubbing with the pestle, the sample of fluorspar with 10 times its weight of powdered silica. After mixing, transfer to the flask, cork the latter and connect up the apparatus as shown. Pour 25 cc. of concentrated sulphuric acid through the funnel *c* (and the glass beads) into the flask *d*. As soon as the acid is poured into the funnel, close the latter with the stopper and guard-tube *b*. Place the flask in a bath of paraffin and heat to about 160° C. for one or two hours.

* This tube serves to remove any sulphuric acid mechanically carried over with the gas.

A slow current of air is passed through the apparatus during this time. The test-tube *f* is then disconnected from the condenser-tube *e*, and its contents together with that of *g* are rinsed into the beaker. Any pasty silicic acid adhering to the sides of *f* must be removed with a stirring-rod, for otherwise some enclosed acid will escape the titration. Add a drop of the lacmoid indicator to the contents of the beaker, and titrate the mixture with the standard alkali. Calculate the percentage of fluorine, as usual, from the weight of sample taken and the number of cubic centimeters of alkali required.

The process depends upon the fact that upon treatment of fluorides with sulphuric acid and silica the following reaction takes place:

$$2CaF_2 + 2H_2SO_4 + SiO_2 = 2CaSO_4 + 2H_2O + SiF_4.$$

Upon coming in contact with water, the silicon fluoride is decomposed into hydrofluosilic and silicic acids:

$$3SiF_4 + 2H_2O = 2H_2SiF_6 + SiO_2.$$

The hydrofluosilicic acid unites with the potassium chloride, forming potassium silicofluoride (insoluble in 50 per cent. alcohol) and hydrochloric acid:

$$H_2SiF_6 + 2KCl = K_2SiF_6 + 2HCl.$$

The hydrochloric acid set free is, of course, titrated, and as $3F_2 = 2HCl$, 1 cc. of N/5 alkali = 0.0114 grams of fluorine.

Sufficient alcohol should always be added to the liquid to be titrated to have at least fifty per cent. present by volume, otherwise the silicofluoride will fail to be completely precipitated, and, as the end-point is reached, will slowly decompose and again cause an acid reaction.

CHAPTER XV.

IRON.

1. In the determination of iron in ores and metallurgical products some modification of either the permanganate or dichromate volumetric method is ordinarily employed. Both methods give exact results, and the choice of the one to use usually depends on either the personal preference of the operator or the convenience of the case in hand.

The following modifications of these methods are employed in my laboratory.

2. **Permanganate Method.**—This method is based on the following reaction:

$$10FeSO_4+2KMnO_4+8H_2SO_4$$
$$=5Fe_2(SO_4)_3+2MnSO_4+K_2SO_4+H_2O.$$

The ferrous salt, which may be either a sulphate or a chloride, is oxidized to the ferric condition by the permanganate, which is itself decomposed and decolorized. When the permanganate in solution is added gradually, its color is continually destroyed as long as any ferrous salt remains, but as soon as the oxidation is complete the addition of more permanganate imparts a permanent pink tint to the liquid. The above equation shows that it requires 316.3 parts of potassium permanganate to oxidize 559 parts of iron from the ferrous to the ferric condition. To make a standard solution, therefore, so that 1 cc. shall equal 1 per cent. of iron, i.e., 0.005 gram when 0.5 gram of ore is taken for assay, we solve the proportion

$$316.3:559=x:0.005.$$

$x=0.00283$. This is the weight of permanganate required in 1 cc., which is equivalent to 2.83 grams per liter. When used for iron determinations, the solution may conveniently be made of approximately this strength. Its exact value is determined by standardization.

3. There are three common methods of standardizing the permanganate solution:

1. By metallic iron or an iron solution of known strength.
2. By a stable ferrous salt.
3. By oxalic acid or an oxalate.

I have found the metallic iron method, when carried out as described below, and the oxalic acid or oxalate method to give practically identical results, and I ordinarily use the latter on account of its simplicity. The stable ferrous salt usually employed is ferrous ammonium sulphate. As ordinarily obtainable it is not a sufficiently reliable basis for the best work.

After making up the permanganate solution it should be allowed to stand at least a day before standardizing, as when first prepared it is constantly changing in strength, owing to oxidation of the organic matter always present in the liquid.

4. To standardize by means of metallic iron, weigh 0.15–0.20 gram of very finely drawn clean iron wire and dissolve in an 8-oz. flask in a mixture of 5 cc. each of strong nitric and hydrochloric acids.

Very finely drawn polished iron wire can be purchased at the supply stores marked with its true percentage of iron, usually about 99.8. Sutton recommends the use of soft "flower-wire," of which the actual percentage of iron may be assumed to be 99.6. Pure electrolytic iron in granulated form is also to be obtained. It has the advantage of dissolving very rapidly in hydrochloric or dilute sulphuric acid, and is said to keep well without oxidizing. When iron wire is used the precaution should be taken to have it perfectly clean and free from rust. If polished and apparently clean, it may simply be drawn through a piece of filter-paper held in the hand to insure freedom from dust and grease, but if at all oxidized it should first be thoroughly cleaned with fine emery-paper or cloth.

5. Having placed the wire in the flask, best in a snug coil, add the acid and warm the mixture gently. The wire will quickly dissolve. Besides effecting a rapid solution of the iron, the aqua regia also serves to oxidize hydrocarbons and other compounds, due to impurities in the wire, that would otherwise consume a little permanganate in the titration. When the wire has dissolved add 5 cc. of strong sulphuric acid and boil over a free flame (manipulating the flask in a holder) until the hydrochloric and nitric acids are expelled and most of the sulphuric also. Allow to cool and add 30 cc. of cold water, 10 cc. of strong hydrochloric acid and 3 cc. of 4% $CuSO_4$ solution.

6. Now add 6 grams of pure granulated zinc (best about 20-mesh). This should be roughly weighed, as all zinc contains a little iron and the error thus introduced must be subsequently determined and allowed for. The iron is reduced from the ferric to the ferrous condition according to the equation

$$2FeCl_3 + Zn = 2FeCl_2 + ZnCl_2.$$

Hydrogen is also liberated by the action of the free acid on the zinc. The copper added precipitates arsenic.

Allow the reaction to continue until the solution is completely decolorized,* heating, if necessary, toward the end if the action is slow. Now add 50 cc. of cold water to the mixture in the flask, and then 10 cc. of strong sulphuric acid. This will dissolve whatever zinc remains and will also supply the solution with a large excess of sulphuric acid. The latter will counteract the otherwise disturbing influence of the hydrochloric acid present during the subsequent titration and secure a sharp end-point. Hydrochloric acid and potassium permanganate mutually deconpose each other with the evolution of chlorine as follows:

$$2KMnO_4 + 16HCl = 2KCl + 2MnCl_2 + 8H_2O + 10Cl.$$

When this takes place during an iron titration, the amount of permanganate consumed is increased, thus producing high results, while the end-point is obscured owing to the continual fading of the pink tinge. If, however, the solution is cold, largely diluted, and contains a proper excess of sulphuric acid, a little hydrochloric acid causes no trouble.†

7. When the zinc is nearly all dissolved the mixture should be filtered. Nearly all zinc contains lead as an impurity, and this remains behind as a finely divided black residue, which would cause trouble by consuming permanganate if not filtered off. As the solution has a tendency to clog a filter and run very slowly it is better to filter through a plug of absorbent cotton placed in a funnel and moistened. Do not use an unnecessarily large wad. Place it in the funnel so that the under side is

---

* If desired, the completion of the reduction may be determined by removing a drop of the liquid with a glass rod and touching it on a porcelain plate with a drop of ammonium thiocyanate solution. Any ferric iron remaining will produce a red tinge.

† It appears to make a very slight change in the standard, but this is of no consequence in technical work, as iron determinations are made the same way.

smooth, with the fibers lying horizontally. If the latter are allowed to string downward they are liable to obstruct the neck of the funnel and cause a slow filtration. The proper arrangement filters rapidly and washes easily. Receive the filtrate in a 4-inch by 5-inch glass battery-jar containing an inch or two of water. Wash the filter and residue well with cold water.

8. Dilute the solution in the battery-jar with cold distilled water to about 700 cc., having previously placed a mark on the jar at that point. The solution is now ready for titration, which should be proceeded with without unnecessary delay, to avoid oxidation of the ferrous iron by the air. Place a piece of white paper under the jar, in order that the color change may be rendered more distinct, and run in the permanganate solution from a burette while stirring the mixture with a glass rod. The color of the permanganate is almost instantly destroyed at first, but as the end-point is approached a less rapid decolorization can easily be detected if the permanganate be added cautiously. Proceed more and more carefully, finally drop by drop, until a very faint permanent pink tinge is obtained. The reaction being now complete, read the burette. Another drop of permanganate should turn the solution decidedly pink.

Before calculating the standard of the permanganate it is necessary to apply a correction by deducting the amount of permanganate solution required to produce the same pink tint in a solution containing no dissolved iron wire, but otherwise of the same volume and condition and containing the same reagents. To determine this correction once for all for the same reagents proceed as follows:

9. Treat 6 grams of the zinc regularly used, in an 8-oz. flask, with a mixture of 10 cc. of strong hydrochloric acid and 25 cc. of water and add gradually a mixture of 10 cc. of strong sulphuric

acid and 20 to 30 cc. of water. When the zinc has all dissolved, dilute with cold water and filter through absorbent cotton as described for the standardization, finally titrating with permanganate precisely as before. Note the amount of permanganate required and deduct this volume from the burette reading of all corresponding iron titrations.

A correction is sometimes made for the color due to the ferric salt present at the end of a titration, but the error from this cause is usually negligible.

10. Having now, in the standardization of the permanganate, taken the final reading of the burette and deducted the correction, there remains only to divide the actual weight of iron in the iron wire taken by the number of cubic centimeters used to find the value of 1 cc. in iron.

Example.—Weight of iron wire taken......	0.1725	gram
Actual weight of iron (99.8%) ..	0.1721	"
Burette reading...............	34.90	cc.
Correction	0.20	"
Corrected reading.............	34.70	cc.

$$0.1721 \div 34.70 = 0.004961.$$

This is the weight of iron in grams to which 1 cc. of the permanganate is equivalent.

On the basis, then, of 0.5 gram of ore being taken for assay, 1 cc. = 0.9922 per cent. of iron.

11. To standardize by means of a ferrous salt, ferrous ammonium sulphate is commonly used. This salt has the composition $FeSO_4(NH_4)_2SO_4+6H_2O$. It contains 14.25 per cent. of iron, although it is frequently assumed to contain exactly one-seventh of its weight of iron. Weigh from 1 to 1.5 grams of the pure salt and dissolve in a battery-jar in 700 cc. of cold distilled water acidified with 10 cc. of strong sulphuric acid. Titrate at once to the usual pink tint. Make a blank test on the plain acidulated water to determine the amount of permanganate required to produce a tint and deduct the correction thus found from the former reading. Calculate the standard as above.

12. To standardize with oxalic acid proceed as described in x, 6, and determine the oxalic acid value of 1 cc. of the permanganate. This value multiplied by 0.8867 will give the iron factor. This is a convenient and accurate method, and is the one usually employed in my laboratory.

13. **Treatment of an Ore.**—Weigh 0.5 gram of the ore and place in an 8-oz. flask. Decompose with acids according to the nature of the sample. It is usually best to begin with 10 cc. of strong hydrochloric acid and warm gently as long as decomposition appears to progress, adding more acid if necessary. If undecomposed sulphides remain, add 5 cc. of strong nitric acid and continue the heating. Note whether the insoluble residue now appears white or discolored. In the latter case endeavor to effect a better decomposition by continued gentle heating with hydrochloric acid. Finally, add to the mixture (which should not be so concentrated as to contain separated salts) about 5 cc. of strong sulphuric acid and heat the flask, supported in a holder, over a free flame until practically all the acid, including the sulphuric, is expelled.* Allow to cool, add 30 cc. of cold water and 10 cc. of strong hydrochloric acid, and the mixture is at once ready for the reduction of the iron with zinc. It is not necessary to get the salts into solution before adding the zinc. They will dissolve very quickly after reduction begins.

---

* Nitric acid may not be entirely expelled by this treatment unless all nitrates are in solution when the sulphuric acid is added.

If hydrochloric acid alone effects a solution of all the iron in the ore, the further treatment with nitric or sulphuric acid is unnecessary. In such a case, in order to have the right amount of acid finally present, boil the solution nearly to dryness and then add 8 cc. more of hydrochloric acid. Dilute to 40 cc. with cold water and proceed with the reduction.

14. Add 6 grams of pure granulated zinc (20-mesh) and reduce the iron precisely as described for the standardization of the permanganate by the metallic iron method (6).

When the reduction is complete, dilute with 50 cc. of cold water and then, with a little caution, add 10 cc. of strong sulphuric acid. Besides a residue of lead from the zinc itself, reducible metals from the ore, such as lead, copper, arsenic, etc., will be precipitated. All this residue must be removed. When the zinc is nearly all dissolved, filter the mixture through a plug of moistened absorbent cotton into a battery-jar, as described in 7, washing the filter well with cold water. Dilute to 700 cc. and titrate at once with permanganate, as described in 8. After reading the burette, deduct the correction previously determined (9) and multiply the actual number of cubic centimeters used for the ore by the percentage value of 1 cc. for iron. This gives the percentage of iron in the sample.

15. **Effect of a Large Amount of Arsenic.**—Ores containing a large amount of arsenic will frequently give low and variable results, and the titration will have a rather quickly fading end-point. Apparently, some arsenic has been left in solution, after the reduction, in the *ic* condition, and this, by subsequently oxidizing iron, brings the results low, as the arsenious acid produced is only very slowly acted upon by the permanganate. The addition of cupric sulphate solution, as prescribed in 5, usually prevents this trouble by precipitating any arsenic in solution. The zinc alone cannot be depended upon to do this.

16. Treatment of Silicates and Other Refractory Substances.—The acid treatment just described will serve excellently for the decomposition of most ores treated at western lead-smelters, but silicates, furnace products, refractory oxides, etc., are frequently encountered that fail to thus yield up all their iron. In the purchase of ores for lead-smelters the actual total iron contents are not, as a rule, absolutely required, but only what is obtained by the acid treatment. In the majority of cases practically all the iron is thus dissolved. What remains is classed with the insoluble residue or "silica." Whenever the total iron is required the material must, of course, be completely decomposed.

17. Silicates.—Silicates, or mixed material containing silicates, may be decomposed as described under SILICA. The nature of the material has to be considered in deciding upon the best course to pursue and the matter of the decomposition is fully explained in the section referred to. An acid solution of the iron is thus finally obtained. If in two portions, as the result of an acid treatment followed by a fusion of the residue, they should be united. When only iron is to be determined it is usually unnecessary to evaporate to dryness to remove silica. Warm the solution in a beaker, add a little bromine water if the iron is not fully peroxidized, and then precipitate the iron as ferric hydroxide with excess of ammonia. Heat to boiling, allow to settle, filter, and wash thoroughly with hot water. With a jet from the wash-bottle transfer the bulk of the washed precipitate from the filter to a beaker, using as little water as possible, and then place the beaker under the funnel. Pour through the filter a hot mixture of 10 cc. of strong hydrochloric acid and 10 cc. of water, so as to dissolve the precipitate still remaining, finally washing the filter with a little hot water. Warm the mixture in the beaker to dissolve all the ferric hydroxide, concentrate if necessary, by boiling, to about 30 cc., and transfer the solution

to an 8-oz. flask. Reduce with zinc and continue in the usual manner, as directed in 14.

18. Decomposition of Silicates by Hydrofluoric Acid.—Sometimes the iron in a silicate, or a silicious residue remaining after acid treatment, can be quickly dissolved as follows:

Treat the substance in a small platinum dish with equal parts of strong pure hydrochloric and hydrofluoric acids. Warm gently until decomposition appears complete, adding more of the acids if necessary, and then add about 3 cc. of dilute sulphuric acid (1:2) and evaporate to white fumes. To the cool residue add a little water and a few drops of hydrochloric acid and warm to effect solution. If the material is a residue, transfer this solution to the flask containing the original filtrate, make to a volume of about 40 cc., reduce with zinc, and finish as usual (14). If the material is the ore itself the operations are similar, but 5 cc. of hydrochloric acid must be added before reduction. When nitric acid has been used in a preliminary treatment it should be expelled before filtration from the residue (13). The solution for reduction should have a volume of about 40 cc. and contain from 5 to 10 cc. of free hydrochloric acid.

19. Refractory Oxides, etc.—Treat 0.5 gram of the finely ground sample with 10–15 cc. of strong hydrochloric acid in an 8-oz. flask. Digest at a very gentle heat, adding more acid, if necessary, as long as the undissolved residue appears to be appreciably attacked. If the ore contains carbonaceous matter, add a few crystals of potassium chlorate to oxidize it. Finally boil or evaporate to a few cubic centimeters and then dilute with 10–15 cc. of water and heat to boiling, adding a few drops of hydrochloric acid, if necessary, to dissolve any separated basic compounds. Filter the solution through a small filter and collect the filtrate in an 8-oz. flask. Transfer the insoluble residue completely to the filter and wash with hot water, but avoid getting too bulky a filtrate for the subsequent reduction. Reserve this solution.

20. Ignite the filter and residue in a small platinum dish until the carbon is burned off and then cool and add about 20 drops of strong sulphuric acid and 2 or 3 cc. of pure strong hydrofluoric acid. Digest at a gentle heat very cautiously to avoid spattering. Finally evaporate to sulphuric acid fumes to expel the hydrofluoric acid, then cool, add a little water and a few drops of hydrochloric acid and warm if necessary to effect solution. If this treatment proves successful the liquid in the dish may be added to the main solution in the flask without filtering, and the reduction and subsequent operations proceeded with as previously described (14).

21. If the mixture of sulphuric and hydrofluoric acids fails to dissolve the residue, proceed as follows: Evaporate to white fumes, add about 0.5 gram of potassium acid sulphate, and heat cautiously, to avoid spattering, until the mass is in quiet fusion and the specks of iron oxide have dissappeared. After cooling, take up by warming with a little water and a few drops of hydrochloric acid. Proceed with the solution as directed above.

22. **Titaniferous Ores.**—In case an ore contains an appreciable amount of titanic acid it becomes impracticable to use zinc as a reducing agent, since the TiO_2 in solution is thereby reduced to Ti_2O_3 (which imparts a purple or blue color to the liquid), and the latter is subsequently oxidized back to TiO_2 again by the permanganate. Thus more permanganate will be required than corresponds to the amount of iron present.

With such titaniferous ores, proceed as described for REFRACTORY OXIDES (19) until the final solution of the ore and residue is obtained in the flask ready for reduction. Now drop into the flask 2 or 3 small spirals of platinum wire to prevent subsequent bumping. Boil the solution, if necessary, so as to reduce its bulk to about 40 cc., and then cautiously add enough ammonia to produce a slight permanent precipitate of ferric hydroxide which persists even after vigorous shaking. Now add 5 cc. of a strong solution of ammonium acid sulphite,* shake well, and then warm the flask gently. As the deep-red color fades, increase the heat gradually to the boiling-point. When quite colorless, add a mixture of 10 cc. of strong sulphuric acid and 20 cc. of water and continue the boiling until all odor of sulphur dioxide has disappeared. Now place the flask in cold water and add cold water until the flask is nearly full. When the solution is quite cold, transfer it to a battery-jar, dilute to 700 cc., and titrate with permanganate as usual (8).

23. Chrome-iron Ore.—Treat as described in XII, 2. After filtering off the chromium the iron is left with the residue on the filter as ferric hydroxide. With as little hot water as possible rinse the residue from the filter into a small beaker. Place the latter under the funnel and dissolve whatever precipitate still adheres to the filter by slowly pouring through the latter a warm mixture of 10 cc. strong hydrochloric acid and 15 cc. water. Heat the mixture in the beaker to dissolve all the ferric hydroxide and then transfer to an 8-oz. flask, reduce with zinc, and determine the iron according to 14.

24. Zimmerman-Reinhardt Method as Modified by Mixer and Dubois.†—This modification of the permanganate method is extensively used in the Lake Superior region for the rapid determination of iron in the oxidized ores of that section.

The following solutions are required:

---

* Ammonium acid sulphite may be made by passing sulphur dioxide into strong ammonia water until the liquid becomes of a yellowish color and smells strongly of sulphur dioxide. An old solution always contains some thiosulphate which will give a precipitate of sulphur with ferric salts.

† Jour. Am. Chem. Soc., XVII, 405.

Stannous Chloride. Dissolve 1 pound of stannous chloride in 1 pound of strong hydrochloric acid (1.2 sp. gr.) to which some water has been added and dilute to 2 liters.

Hydrochloric Acid of 1.1 *Sp. Gr.* Mix equal volumes of the strong acid (1.2 sp. gr.) and water.

Mercuric Chloride. Make a saturated solution in hot water, allow to cool and crystallize and then filter.

Manganese Sulphate. Dissolve 160 grams in water and dilute to 1750 cc. To this add 330 cc. of phosphoric acid syrup of 1.7 sp. gr. and 320 cc. of sulphuric acid, 1.84 sp. gr. This solution is to obviate the deleterious action of liberated chlorine when potassium permanganate is added to a hydrochloric acid solution. The phosphoric acid allows the formation of iron phosphate, which, being nearly colorless, renders the end-reaction more distinct.

25. Treatment of an Ore.—Treat 0.5 gram in a small covered beaker with 2.5 cc. of the stannous chloride solution and 10 to 15 cc. of the dilute hydrochloric acid. Boil the mixture very gently on an iron plate until the ore is completely decomposed. For ores running less than 55 per cent. of iron it is advisable to use a little less stannous chloride. The solution of the ore is usually very rapid, requiring only a few minutes.

When the ore is dissolved, run a few drops of stannous chloride from a burette into the hot solution until all the iron is reduced to the ferrous state, as indicated by the disappearance of the greenish-yellow color. If an excess of stannous chloride has been originally added, it is of course unnecessary to add more. In any case, to avoid too great an excess of stannous chloride, which is undesirable, it is advisable to add a few drops of potassium permanganate solution to the reduced mixture to once more slightly oxidize the solution. The solution, in its slightly oxidized condition, should be kept warm until ready to titrate and then the final reduction made with a drop or two of stannous chloride, avoiding unnecessary excess. Now wash down the sides of the beaker and add, while stirring, 5 cc. of the mercuric chloride solution to take up the excess of stannous chloride. Have ready a 500-cc. beaker containing 6–8 cc. of the manganese sulphate solution and about 400 cc. of cold water. Wash the iron solution into this and titrate at once with potassium permanganate in the usual way.

Ores containing organic matter, some magnetites, and pyritous ores require the usual precautions. With ores containing very large amounts of organic matter, it is generally most advantageous to burn off directly and follow the regular method. Ores containing small amounts of organic matter and slightly pyritous ores are dissolved in hydrochloric acid and oxidized with potassium chlorate, after which the regular method is pursued. Heavy sulphides may be treated with nitric acid as described in **31**, and the hydrochloric acid solution eventually obtained treated as above. Magnetites should be ground very fine in an agate mortar and then, if unable to effect complete solution, either the ore or the insoluble residue may be treated as described in **21**.

26. Dichromate Method.*—The following solutions are required:

Stannous Chloride. This should be strongly acid and contain about 15 grams of tin and 350 cc. of strong hydrochloric acid to the liter. It may be made by dissolving the tin in the acid by the aid of heat and diluting. It is usually more convenient to prepare it from the crystallized stannous chloride as follows: Dissolve 14.5 grams of the crystals in 165 cc. of strong hydrochloric acid and dilute to 500 cc. Keep in a half-liter bottle in which a stick of pure tin is placed to prevent oxidation.

* The method ordinarily employed for ores in my laboratory.

One cubic centimeter of this solution will reduce about 0.015 gram of iron from the ferric to the ferrous condition. It will naturally become somewhat stronger as the stick of tin gradually dissolves in the acid liquid.

Mercuric Chloride. Use a saturated solution and keep an excess of the crystals in the bottle. Such a solution will contain at least 60 grams of mercuric chloride to the liter. About 1.2 cc. of this solution will oxidize the tin in 1 cc. of the above stannous chloride solution, at its original strength, to the stannic condition.

Potassium Ferricyanide. This solution should be dilute, say 0.1 gram in 15 cc. of water. The exact strength is immaterial. It is best made frequently, or when required, in small quantity, as the solution does not keep indefinitely.

27. **Standard Potassium Dichromate.** — This should contain 4.39 grams of the pure salt per liter. On the basis of 0.5 gram of ore being taken for assay, 1 cc. of a solution of exactly this strength will equal 1 per cent. of iron. This is shown by the equation

$$6FeCl_2 + K_2Cr_2O_7 + 14HCl = 6FeCl_3 + 2KCl + 2CrCl_3 + 7H_2O.$$

Thus 1 molecule of potassium dichromate oxidizes 6 molecules of ferrous chloride to ferric chloride, or, 1 molecule of $K_2Cr_2O_7 = 6Fe$. This corresponds to 294.5 parts of $K_2Cr_2O_7$ to 335.4 parts of Fe. Hence 1 per cent., or 0.005 gram of iron when 0.5 gram of ore is taken for assay, requires 0.00439 gram of $K_2Cr_2O_7$. If this amount is contained in 1 cc. then the liter will contain 4.39 grams of $K_2Cr_2O_7$.

28. Standardize the dichromate solution as follows: Weigh carefully 0.15–0.20 gram of the purest iron wire obtainable (see remarks in 4 relative to the iron used for standardizing) and dissolve it in an 8-oz. flask in a mixture of 5 cc. each of strong

nitric and hydrochloric acids. When the wire has dissolved add 5 cc. of strong sulphuric acid and boil over a free flame (manipulating the flask in a holder) until the hydrochloric and nitric acids are expelled and most of the sulphuric also. Before adding the sulphuric acid see that all salts are in solution (if necessary adding more hydrochloric acid), otherwise separated nitrates may not be completely decomposed. Allow to cool, add about 30 cc. of water and 5 cc. of strong hydrochloric acid and heat the mixture gently until solution is complete.

29. To the hot solution now add the stannous chloride solution cautiously until decolorization is complete, avoiding more than a slight excess. Transfer the reduced solution to a large beaker, washing out the flask with cold water. Add about 10 cc. of the mercuric chloride solution, pouring it in rapidly while stirring the mixture. If only a slight excess of stannous chloride was used, the 10 cc. of mercuric chloride will be ample and a white precipitate will be produced, but if the excess of stannous chloride was large the precipitate may become gray or black from the separation of metallic mercury. This discoloration should be watched for carefully, and if the faintest trace of it appears add an abundance of the mercuric chloride solution at once. If this restores the pure white color the test may proceed, otherwise it is spoiled and must be begun anew.

The reactions involved are, first,

$$2FeCl_3+SnCl_2=2FeCl_2+SnCl_4.$$

The stannous chloride thus reduces the ferric to ferrous chloride and becomes itself oxidized to stannic chloride. The excess of stannous chloride added, when treated with sufficient mercuric chloride, is oxidized to stannic chloride as follows:

$$SnCl_2+2HgCl_2=SnCl_4+Hg_2Cl_2.$$

Mercurous chloride is produced, which is a white precipitate. If the excess of stannous chloride is large, however, so that the mercuric chloride added is insufficient for the last reaction, the following takes place:

$$SnCl_2 + HgCl_2 = SnCl_4 + Hg.$$

The finely divided metallic mercury discolors the liquid. If more mercuric chloride be added at once, the mercury may frequently be taken into combination again:

$$Hg + HgCl_2 = Hg_2Cl_2.$$

The liquid therefore contains as a final result, ferrous chloride, stannic chloride, mercuric chloride, and a precipitate of white mercurous chloride. The only one of these compounds affected by the dichromate in the subsequent titration is the ferrous chloride.

30. The mixture in the beaker is now ready for the titration, which should be proceeded with without delay, since after the mercuric chloride has been added there is nothing to prevent oxidation of the ferrous solution. Have ready a glazed porcelain tile either with or without depressions, a beaker or glass full of water in which is placed a glass rod, and also the test solution of potassium ferricyanide. Run in the dichromate solution from the burette while stirring the iron solution, and from time to time place a drop of the latter on the porcelain and touch it with a drop of the ferricyanide. For the latter purpose use the rod placed in the glass of water, and after each test return it to the glass and it will thus be kept sufficiently washed. As long as a large amount of iron still remains in the ferrous condition the ferricyanide will produce intensely blue tests, and, in order not to be deceived in the matter, endeavor to add a sufficiently large drop of the ferricyanide each time to combine and produce a blue color with all the ferrous chloride present in the test drop. Continue to thus run in the dichromate solution until the color of the tests becomes fainter. Proceed then with more caution,

finally drop by drop, until a test is obtained that fails to show a blue tint after waiting half a minute. In making the final tests use less and less of the ferricyanide, as the delicacy of the reaction depends upon having just about enough ferricyanide present to combine with the ferrous chloride still remaining, an excess producing an abnormal color that obscures the true end-point. Finally read the number of cubic centimeters used, and, from the weight of iron taken, calculate the value of 1 cc. in iron.

Example.—Suppose 0.17 gram of iron wire, actually containing 0.1697 gram of iron, were taken, and that 33.8 cc. of dichromate solution were used. Then the value of 1 cc. is 0.1697 ÷ 33.8, or 0.005021 gram of iron.

When 0.5 gram of ore is taken for assay, 1 cc. of the dichromate solution will equal 1.004 per cent. of iron.

31. Treatment of an Ore.—Take 0.5 gram of the ore. The method of decomposition will depend upon the nature of the sample. Whenever possible hydrochloric acid alone should be used. When the sample appears to be more or less of an oxidized nature always try hydrochloric acid first, using 10 cc. and warming the mixture in an 8-oz. flask. Take all the time and acid necessary, but do not boil violently, as this weakens the acid and fails to effect as rapid a solution as a gentler heat. If undecomposed sulphides remain, add 1–2 cc. of strong nitric acid and continue the heating. The final insoluble residue should be clean and white. If hydrochloric acid alone effects a complete decomposition, boil the solution finally to pastiness, add about 30 cc. of water and then 5 cc. of strong hydrochloric acid. These last operations are simply to insure about the proper proportion of acid present for the succeeding steps. If nitric acid has been used in the decomposition, first see that all soluble matter is still held in solution (otherwise adding more hydrochloric acid), then add 5 cc. of strong sulphuric acid and boil, best over a free flame, as nearly as possible to dryness, so as to expel most of the sulphuric acid. It is very difficult, if not impossible, to completely decompose solid nitrates, that have separated from the solution by concentration, by boiling with sulphuric acid. After cooling, add about 75 cc. of water and 5 cc. of strong hydrochloric acid. It is unnecessary to heat the mixture to effect solution of the salts at this stage.

Observe that whatever method of decomposition is employed there is finally obtained a solution or mixture containing about 5 cc. of free hydrochloric acid.

32. To the mixture in the flask add about 0.2 gram of pure cupric sulphate and see that it dissolves, or better, an equivalent amount of strong solution. Now add about 20 grams of pure, finely granulated test-lead, and boil gently to effect the reduction of the iron to the ferrous condition. The copper precipitates on the lead and prevents it from cohering in lumps. If the ore already contains sufficient copper it is of course unnecessary to add more. Avoid adding the lead to a boiling-hot solution or it will probably produce a sudden evolution of steam and overflow the flask. The iron is usually entirely reduced by about five minutes' gentle boiling. It is best to boil for a few minutes after the liquid has become completely decolorized. The copper is thus all precipitated and any arsenic or antimony also removed from the solution. Prolonged boiling causes the formation of too much lead chloride, which separates out and renders the solution difficult to filter. Filter boiling-hot, bringing the excess of lead upon the filter, and wash thoroughly with hot water. Receive the filtrate in a large beaker in which 5 cc. of strong hydrochloric acid and 2 cc. of the prepared stannous chloride solution (26) have previously been placed. The acid is simply to insure sufficient acidity and the stannous chloride is to reduce any ferric chloride that may possibly be present and keep it

reduced. Stir the mixture in the beaker after filtration. As it contains a little stannous chloride the ferrous chloride will not become oxidized by standing a short time at this stage.

33. The burette and other arrangements being ready (as described in 30), add 10 cc. or mercuric chloride solution (26) to the liquid in the beaker and titrate as described for the standardization of the dichromate solution (30). From the number of cubic centimeters of dichromate solution used and the known value in iron of 1 cc., calculate the percentage of iron in the ore.

34. **Silicates, Refractory Oxides, etc.**—Decompose these substances as described either in 17, 18, 19 or 23, so as to finally obtain the iron in hydrochloric acid solution containing about 5 cc. of the strong acid. Proceed with this solution as described in 32.

35. **Standard Methods for the Analysis of Iron.***—

Determination of Silicon.

Weigh 1 gram of sample, add 30 cc. nitric acid (sp.gr. 1.13); then 5 cc. sulphuric acid (conc.). Evaporate on hot plate until all fumes are driven off. Take up in water and boil until all ferric sulphate is dissolved. Filter on an ashless filter, with or without suction pump, using a cone. Wash once with hot water, once with hydrochloric acid, and three or four times with hot water. Ignite, weigh, and evaporate with a few drops of sulphuric acid and 4 or 5 cc. of hydrofluoric acid. Ignite slowly and weigh. Multiply the difference in weight by .4702, which equals the per cent. of silicon.

* 1907 Report of the Committee of the American Foundrymen's Association.

Determination of Sulphur.

Dissolve slowly a 3-gram sample of drillings in concentrated nitric acid in a platinum dish covered with an inverted watch glass. After the iron is completely dissolved, add 2 grams of potassium nitrate, evaporate to dryness and ignite over an alcohol lamp at red heat. Add 50 cc. of a 1 per cent. solution of sodium carbonate, boil for a few minutes, filter, using a little paper pulp in the filter if desired, and wash with a hot 1 per cent. sodium carbonate solution. Acidify the filtrate with hydrochloric acid, evaporate to dryness, take up with 50 cc., of water and 2 cc. of concentrated hydrochloric acid, filter, wash, and after diluting the filtrate to about 100 cc., boil and precipitate with barium chloride. Filter, wash well with hot water, ignite and weigh as barium sulphate, which contains 13.733 per cent. of sulphur.

Determination of Phosphorus.

Dissolve 2 grams sample in 50 cc. nitric acid (sp. gr. 1.13),. add 10 cc. hydrochloric acid and evaporate to dryness. In case the sample contains a fairly high percentage of phosphorus it is better to use half the above quantities. Bake until free from acid, redissolving in 25 to 30 cc. of concentrated hydrochloric acid, dilute to about 60 cc., filter and wash. Evaporate to about 25 cc., add 20 cc. concentrated nitric acid, evaporate until a film begins to form, add 30 cc. of nitric acid (sp. gr. 1.20) and again evaporate until a film begins to form. Dilute to about 150 cc. with hot water and allow it to cool. When the solution is between 70° and 80° C., add 50 cc. of molybdate solution. Agitate the solution a few minutes, then filter on a tared Gooch crucible having a paper disc at the bottom. Wash three times with a 3 per cent. nitric acid solution and twice with alcohol. Dry at 100° to 105° C., to constant weight. The weight multiplied by 0.0163 equals the per cent. of phosphorus in a 1 gram sample.

To make the molybdate solution add 100 grams molybdic acid to 250 cc. water, and to this add 150 cc. ammonia, then stir until all is dissolved and add 65 cc. nitric acid (sp. gr. 1.42). Make another solution by adding 400 cc. concentrated nitric acid to 1,100 cc. water, and when the solutions are cool, pour the first slowly into the second with constant stirring and add a couple of drops of ammonium phosphate.

Determination of Manganese.

Dissolve 1.10 grams of drillings in 25 cc. nitric acid (sp. gr. 1.13), filter into an Erlenmeyer flask and wash with 30 cc. of the same acid. Then cool and add about 0.5 gram of sodium bismuthate until a permanent pink color forms. Heat until the color has disappeared, with or without the precipitation of manganese dioxide, and then add either sulphurous acid or a solution of ferrous sulphate until the solution is clear. Heat until all nitrous oxide fumes have been driven off, cool to about 15° C.; add an excess of sodium bismuthate — about 1 gram — and agitate for two or three minutes. Add 50 cc. water containing 30 cc. nitric acid to the liter, filter on an asbestos filter into an Erlenmeyer flask, and wash with 50 to 100 cc. of the nitric acid solution. Run in an excess of ferrous sulphate and titrate back with potassium permanganate solution of equal strength. Each cubic centimeter of N/10 ferrous sulphate used is equal to 0.10 per cent. of manganese.

Determination of Total Carbon.

This determination requires considerable apparatus; so in view of putting as many obstacles out of the way of its general adoption, in cases of dispute your committee has left optional several points which were felt to bring no chance of error into the method.

The train shall consist of a pre-heating furnace, containing copper oxide (Option No. 1) followed by caustic potash (sp. gr. 1.20), then calcium chloride, following which shall be the combustion furnace in which either a porcelain or platinum tube may be used (Option No. 2). The tube shall contain 4 or 5 inches of copper oxide between plugs of platinum gauze, the plug to the rear of the tube to be at about the point where the tube extends from the furnace. A roll of silver foil about 2 inches long shall be placed in the tube after the last plug of platinum gauze. The train after the combustion tube shall be anhydrous cupric sulphate, anhydrous cuprous chloride, calcium chloride, and the absorption bulb of potassium hydrate (sp. gr. 1.27) with prolong filled with calcium chloride. A calcium chloride tube attached to the aspirator bottle shall be connected to the prolong.

In this method a single potash bulb shall be used, a second bulb as sometimes used for a counterpoise being more liable to introduce error than correct error in weight of the bulb in use, due to change of temperature or moisture in the atmosphere.

The operation shall be as follows: To 1 gram of well mixed drillings add 100 cc. of potassium copper chloride solution and 7.5 cc. of hydrochloric acid (conc.). As soon as dissolved, as shown by the disappearance of all copper, filter on previously washed and ignited asbestos. Wash thoroughly the beaker in which the solution was made with 20 cc. of dilute hydrochloric acid (1 : 1), pour this on the filter and wash the carbon out of the beaker by means of a wash bottle containing dilute hydrochloric acid (1 : 1) and then wash with warm water until all the acid is washed out of the filter. Dry the carbon at a temperature between 95° and 100° C.

Before using the apparatus a blank shall be run and if the bulb does not gain in weight more than 0.5 milligram, put the dried filter into the ignition tube and heat the pre-heating furnace and the part of the combustion furnace containing the copper oxide. After this is heated start the aspiration of oxygen or air at the rate of 3 bubbles per second, to show in the potash bulb. Continue slowly heating the combustion tube by turning on two burners at a time, and continue the combustion for 30 minutes if air is used; 20 minutes if oxygen is used. (The Shimer crucible is to be heated with a blast lamp for the same length of time.)

When the ignition is finished turn off the gas supply gradually so as to allow the combustion tube to cool off slowly and then shut off the oxygen supply and aspirate with air for 10 minutes. Detach the potash bulb and prolong, close the ends with rubber caps and allow it to stand for 5 minutes, then weigh. The increase in weight multiplied by 0.27273 equals the percentage of carbon.

The potassium copper chloride shall be made by dissolving 1 pound of the salt in 1 liter of water and filtering through an asbestos filter.

Option No. 1. — While a pre-heater is greatly to be desired, as only a small percentage of laboratories at present use them it was decided not to make the use of one essential to this method; subtraction of the weight of the blank to a great extent eliminating any error which might arise from not using a pre-heater.

Option No. 2. — The Shimer and similar crucibles are largely used as combustion furnaces, and for this reason it was decided to make optional the use of either the tube furnace or one of the standard crucibles. In case the crucible is used it shall be followed by a copper tube 3/16 inch inside diameter and 10 inches long, with its ends cooled by water jackets. In the center of the tube shall be placed a disk of platinum gauze, and for

3 or 4 inches in the side towards the crucible shall be silver foil and for the same distance on the other side shall be copper oxide. The ends shall be plugged with glass wool, and the tube heated with a fish tail burner before the aspiration of air is started.

Graphite.

Dissolve 1 gram of sample in 35 cc. of nitric acid (sp. gr. 1.13), filter on asbestos, wash with hot water, then with potassium hydrate (sp. gr. 1.1), and finally with hot water. The graphite is then ignited as specified in the determination of total carbon.

CHAPTER XVI.

LEAD.

MANY methods have been proposed for the technical determination of lead in ores, etc. The following method, which is the result of much patient experiment, is in daily use in my laboratory. I find it more generally satisfactory than any other.

1. **Chromate Method.**—Prepare the following solutions:

Extraction Solution. Make a cold saturated solution of commercial sodium acetate in distilled water and filter it. Mix 1 volume with two volumes of distilled water and add to the mixture 30 cc. of 80 per cent. acetic acid per liter. Used nearly boiling in a wash-bottle.

Hydrochloric Acid Mixture. Make a cold saturated solution of table salt in distilled water and filter it. To 1 liter of the filtrate add 150 cc. of distilled water and 100 cc. of strong hydrochloric acid. Used cold in a wash-bottle.

Potassium Dichromate. A cold saturated solution of the powdered commercial salt in distilled water. Filter, or allow to settle.

Starch Solution. See XIII, 2.

2. **Treatment of an Ore.**—Weigh 0.5 gram of the ore into an 8-oz. flask. It is usually best to begin the treatment with 20 cc. of strong hydrochloric acid and heat gently until all iron oxide, etc., is in solution. This will also decompose galena and expel the hydrogen sulphide. If sulphides remain that resist hydrochloric acid, add 5 cc. of strong nitric acid and continue

the heating. Finally, if necessary, add 5–10 cc. more hydrochloric acid and again heat, to bring all the lead chloride into solution.

3. Decomposition and solution having been satisfactorily effected, add 5 cc. of strong sulphuric acid and boil, finally, over a free flame, until the white fumes are coming off copiously. Cool, add about 30 cc. of water and heat to boiling. Allow to stand, hot, until the anhydrous ferric sulphate usually present has dissolved. Now add 10 cc. of alcohol (either grain or wood) and then cool to room temperature, or cooler, and filter through a 9-cm. filter. (The alcohol is not absolutely essential except for the low percentages, say below ten per cent. lead.) Wash the precipitate with cold dilute (1–10) sulphuric acid at least four times.

4.* Now open the filter carefully and spread it in the funnel. Wash the precipitate from it into the flask again with a jet of hot water, using as little as possible. Wash every trace of residue from the filter with the extraction solution, nearly boiling. Add sufficient additional hot extraction solution to the mixture in the flask to complete the solution of all the lead sulphate. If apparently necessary, which is seldom, heat to boiling and filter, washing with extraction solution. Dilute the solution, or the filtrate, to 150 cc. with hot water. It is best to have a similar flask, with a mark on it for comparison, so as to always obtain, approximately, the right volume. Heat the mixture to boiling and see that all lead sulphate is in solution. The amount of gangue in the unfiltered mixture is usually small and of no consequence. To the boiling solution add, with a pipette, 10 cc. of the prepared potassium dichromate solution. Boil the mixture very gently for seven minutes. About this length of time is necessary to insure

* It is safer in all cases, even in the absence of barium, to proceed as described in **8**. The method is longer, but the complete extraction from the filter more certain.

always the same constitution of the lead chromate; somewhat longer does no harm. Now filter through an 11-cm. filter. Wash out the flask with hot water and then wash the filter and precipitate ten times with a hot dilute solution of sodium acetate (50 cc. of a cold saturated solution of the commercial salt diluted to 1 liter), using about 5–6 cc. each time. Now again place the clean flask under the funnel, and, with a jet of the cold hydrochloric acid mixture, stir up and dissolve the precipitate on the filter and continue washing with the same mixture until all the residue and every trace of color are removed from the filter. Use at least 50 cc. of the mixture in any case, in order that the subsequent reactions may have the proper conditions. Wash the filter finally with cold water, or at once dilute the filtrate until the flask is half full. Either one of two procedures is now to be followed, according to the amount of lead apparently present.

5. *With not over twenty per cent. lead.* Add 2 cc. of a fifty per cent. solution of potassium iodide, mix and titrate at once (to avoid possible loss of iodine by volatilization) with a standard sodium thiosulphate solution until the brown color becomes faint; then add sufficient starch solution to produce a strong blue color and finish the titration very slowly, finally drop by drop, until the solution becomes a clear pale green with no tinge of blue. The end-point is very sharp and care must be exercised or it may easily be passed. It is best to have a white surface under the flask.

6. *With over twenty per cent. lead.* Pour all but about 25 cc. of the liquid in the flask into a beaker. To the solution still remaining, add about 25 cc. of the hydrochloric acid mixture and 2 cc. of the above potassium iodide solution. Mix and titrate at once with the standard thiosulphate until the brown color has become faint. Now add a portion of the solution reserved in the beaker and again titrate to a faint brown color. Continue

thus until the last of the reserved solution has been rinsed into the flask; then finish the titration as in 5. In pouring from the beaker it is best to use a glass rod, so as to avoid having to wash the lip each time and unnecessarily increasing the volume of liquid in the flask, otherwise the latter may become inconveniently full.

7. Standardize the thiosulphate solution on pure lead. The thiosulphate solution used in the iodide copper method and containing about 19.5 grams of the crystallized salt per liter is satisfactory for lead. Dissolve about 0.2 gram of pure lead foil in a little 1:2 nitric acid contained in an 8-oz. flask, and boil the solution to complete dryness. Cool, add about 30 cc. of water, then 5 cc. of strong sulphuric acid. Boil the mixture a short time, add a little water to replace that lost by boiling, then 10 cc. of alcohol, and cool to room temperature or cooler. Filter the lead sulphate and continue as described for ores (4). Titrate as described in 6.

One cubic centimeter of the thiosulphate will be found to equal a little more than 0.005 gram of lead, or something over one per cent. on the basis of 0.5 gram of ore taken for assay The copper value of the thiosulphate, multiplied by 1.078, will give a close approximation to the lead value.

Notes. The final solution is not a very good solvent for iodine, and therefore, if much is liberated at once, there is danger of loss by volatilization. To avoid this, the procedure described in 6 for the higher percentages of lead has been found the best.

The amount of potassium iodide prescribed is much more than theoretically required in any case, but it is unsafe to ever use less, as the reactions are then liable to proceed too slowly.

Testing the completness of the extraction is rarely necessary, as the solution of all the lead sulphate is usually easily seen.

Prepare only a wash-bottleful of the dilute sodium acetate

wash at a time, as, unless the solution is heated frequently, a fungus growth develops that will clog the filter.

If the end-point be passed in titrating, it may be "brought back" with a dilute solution of potassium dichromate or permanganate, of which the comparative value need not be known. Simply run in from a burette a measured number of drops until the blue color returns; titrate once more and take reading. Now again run in the same number of drops of the correction solution as before and again finish the thiosulphate titration and read the burette. Subtract the difference between the two readings from the first one to obtain the true end-point for the lead.

If, in titrating by portions, as in 6, the end-point be slightly passed with any of the portions, no appreciable error will be introduced if more of the reserved solution be quickly added, as the thiosulphate solution is decomposed only very slowly by the acid mixture.

The constitution of the lead chromate is a vital point. It depends upon the various conditions of heat, acidity, etc. By operating as described, differences in constitution are minimized and a slightly acid chromate of practically constant composition is obtained. The normal chromate, under the conditions obtaining, is difficult to maitain as such.

Bismuth in small amount does not interfere. With several per cent. present, some may remain as basic sulphate with the lead sulphate and produce high results. If bismuth is suspected it may be removed just before filtering the lead chromate by adding about 2 grams of citric acid dissolved in a little hot water. Any bismuth chromate will go into solution at once. Filter without delay.

8. Ores Containing Barium.—When a lead ore contains barium it is frequently very difficult to extract the lead completely from the mixed sulphates by the usual procedure. In such a case, either before or after attempting the usual extraction, drop the filter and residue into an 8-oz. flask, add 5–10 cc. of strong hydrochloric acid and boil to pastiness, almost to dryness, the filter being converted into a pulp. Avoid going so far as to burn the filter or dry it on to the flask. Now add about 25 cc. of the usual acetate solution (or the acetate filtrate), boil and filter. Wash thoroughly, first with the hot acetate solution, then with hot water. Proceed with the filtrate in the usual manner.

9. **Alexander's Method, Modified.**—Alexander's method, or or some modification of it, is the one most commonly employed in western smelting works. The principal objections to it that I have noted are the difficulty of accurately determining very low percentages of lead, on account of the slowness of the reactions in such cases, and the occasional failure to obtain checking results or satisfactory end-points with the higher percentages, for no apparent reason. The following modification was long used in my laboratory:

Treat 0.5 gram of the ore precisely as described in 2 and 3 and extract the lead sulphate as described in 4. (If it is desired to test the completness of the extraction, use hydrogen sulphide water, instead of potassium dichromate, and subsequently add any tested washings showing lead to the main filtrate. If ammonium sulphide is used for testing, a discoloration might be obtained that was due to iron, which is frequently present in this filtrate.) Add to the filtrate in the flask an excess of ammonium sulphide. It is a good plan, although not essential, to first add a little ammonia to neutralize the acid and prevent separation of sulphur. Boil for a moment to coagulate the lead sulphide and then filter and wash with hot water. These operations entirely remove any calcium sulphate.

10. The lead sulphide is still almost certain to contain a

little iron. This iron is originally retained by the lead sulphate. If allowed to remain it tends to obscure the end-point of the subsequent titration by producing a color with the tannic acid used for testing. It is best, therefore, to remove any possible iron in every case as follows: Pour through the precipitate on the filter a mixture of 5 cc. of the dilute (1:10) sulphuric acid and 15 cc. of strong hydrogen sulphide water, and then wash with cold water.

11. Now drop filter and contents into the flask and add 5 cc. of strong hydrochloric acid. Boil the mixture, best by manipulating the flask over a small free flame to prevent bumping, to convert the lead to chloride and expel hydrogen sulphide, the filter meanwhile becoming well disintegrated. If 5 cc. of acid prove insufficient to entirely decompose the lead sulphide, add a little more, but avoid using a large excess. In any case boil off half or more of the acid and then, to the hot mixture, add 2–3 drops of strong nitric acid to oxidize any unexpelled hydrogen sulphide. Now add 25 cc. of cold water, then a few drops of litmus solution as an indicator, and then cautiously add ammonia in very slight excess. Finally, make distinctly acid with acetic acid.

12. Heat the mixture in the flask to boiling, dilute to about 200 cc. with boiling-hot water, and titrate with the standard ammonium molybdate solution as follows: Pour about two-thirds of the hot lead solution into a large beaker and run the molybdate solution into it from a burette until a drop from the beaker, when placed on a glazed porcelain plate and touched with a drop of a solution of tannic acid (about 0.1 gram dissolved in 20 cc. of water), gives a brown or yellow tinge. Now add more of the lead solution from the flask and continue the titration until the end-point is again passed. Continue thus to approach the true end-point, using more caution each time. Finally, when only a few cubic centimeters remain in the flask, pour the entire

mixture in the beaker into the flask and then back into the beaker again and finish the titration 2 drops at a time. When the final yellow tinge is obtained, some of the immediately preceding tests may have developed a tinge also. From the reading of the burette deduct the volume of 2 drops for each test thus showing a color. Multiply the corrected reading by the percentage value of 1 cc. of the molybdate solution in lead to obtain the percentage of lead in the ore.

13. **Standard Molybdate Solution.**—This should contain about 4.74 grams of ammonium molybdate per liter, in order that when 0.5 gram of ore is taken for assay 1 cc. shall equal about 1 per cent. of lead.

Standardize as follows: Weigh carefully about 0.2 gram of pure lead foil and dissolve in an 8-oz. flask by warming with a mixture of 2 cc. of strong nitric acid and 4 cc. of water. When dissolved, boil nearly or quite to dryness, add about 30 cc. of water, and see that all the lead nitrate dissolves. Now add 5 cc. of strong sulphuric acid, boil the mixture a moment, cool to room temperature, and allow to stand and settle a short time. Filter and wash with dilute (1:10) sulphuric acid. Proceed with the filtered lead sulphate precisely as described for the assay of an ore (in 9 et seq.), except that the purification of the lead sulphide described in 10 may be omitted. I formerly dissolved the lead sulphate directly in a hot solution of ammonium chloride and acetate and titrated at once. It was observed, however, that the large amount of ammonium salts necessary to effect complete solution of the lead sulphate hindered the separation of lead as molybdate during the titration, and the end-point was not as sharp as that obtained by the method described.

Divide the weight of lead taken by the number of cubic centimeters of molybdate solution used. This will give the weight of lead corresponding to 1 cc. of molybdate. From this figure calculate the percentage value of 1 cc. on the basis of 0.5 gram of ore taken for assay. 1 cc. of the molybdate solution should thus equal about 0.005 gram of lead, or about 1 per cent.

14. Shorter Method for Ores containing Little or No Calcium. — Alexander's original method did not sufficiently provide for the presence of calcium, which is a frequent constituent of lead ores. It was partially to avoid this source of error that the method was modified as above. Calcium forms a molybdate which is more or less insoluble under the conditions of the titration and tends to raise the results in a rather irregular manner. When it is known to be absent, or present only in small amount (as may frequently be noted from the appearance of the lead sulphate on the filter), the above process may be shortened as follows:

15. Begin as usual and proceed until the washed lead sulphate precipitate is obtained on the filter. Place the precipitate and filter in the original flask, add 10 cc. of strong hydrochloric acid and boil until the filter is well disintegrated, then add 15 cc. of strong hydrochloric acid, 25 cc. of cold water and 25 cc. of strong ammonia water. Now color with a little litmus solution, make slightly alkaline with ammonia if not already so, and then make distinctly acid with strong acetic acid. Heat to boiling and see that the lead sulphate is entirely dissolved, then dilute to about 200 cc. with boiling-hot water and titrate as previously described (12).

16. Even in the absence of calcium this method is liable to be unsatisfactory. The lead sulphate precipitate will frequently retain iron that cannot be dissolved or washed out, but which finally goes into solution with the lead sulphate and gives a color with tannic acid on the test-plate. The large amount of salts present also obscures the end-point and hinders the precipitation of the lead molybdate. This is especially noticeable with

very low grade ores. Except, therefore, for hasty approximate results, it is generally best to use the somewhat longer method first given.

17. Special Treatment for Roasted Products. — Roasted products are liable to have some of the lead in the form of a silicate that resists the ordinary treatment with acids. The following method of procedure has been found successful in such cases:

Treat the material in the usual way, with the addition of about 3 cc. of strong hydrofluoric acid. Ordinarily, this is all that is necessary. If there is much carbonaceous matter the mixture should be again filtered after treatment with the extraction solution. In doubtful cases, reserve this filtrate, hot, instead of proceeding with it, until the insoluble residue on the filter can be further treated as follows:

Wash the filter and residue well with hot water (to remove sodium acetate) and then ignite them at a low temperature in a platinum dish. The amount of lead present is usually too small to occasion any danger to the dish. When the paper is burned off, cool, add a little ammonium nitrate and ignite gently once more. When cool, add about 3 cc. of hydrochloric acid, 3 cc. or more of pure strong hydrofluoric acid, and a few drops of strong sulphuric acid. Evaporate cautiously until the sulphuric acid is fuming strongly, then cool, dilute with a little water and heat to boiling. Keep hot until any anhydrous ferric sulphate has entirely dissolved, then cool to room temperature and filter, washing with the usual dilute sulphuric acid. Extract the lead sulphate on the filter with hot extraction solution, receiving the filtrate in the flask containing the main portion. Finish from this point in the regular way, according to the method employed.

18. Determination of Lead in the Fire Assay Button.—Hammer or roll the button out thin and weigh 0.250 gram. Dissolve by warming in an 8-oz. flask with a little 1:2 nitric acid. When dissolved, boil nearly to dryness, add about 30 cc. of water and see that all the lead nitrate dissolves. Some lead sulphate, due to sulphur in the lead button, may remain insoluble. Proceed from this point as described in 7 or 13, according to the method adopted. Calculate the percentage of lead in the button and correct the fire assay accordingly.

CHAPTER XVII.

MAGNESIUM.

1. **Method for Ores, etc.**—Treat 0.5 gram of the ore in an 8-oz. flask with 10 cc. of strong hydrochloric acid, heating gently, avoiding boiling, to dissolve oxides, etc. Then, if sulphides are also present, add 5 cc. of strong nitric acid and continue the gentle heating until they are decomposed. Now boil to dryness to expel the acids (and render any separated silica insoluble). Warm the residue with 25 cc. of water and 1 or 2 cc. of strong hydrochloric acid, to effect solution of the salts, then transfer to a beaker, dilute to about 150 cc. with cold water and pass in hydrogen sulphide gas to remove copper, lead, etc. Filter, washing with dilute hydrogen sulphide water, and receive the filtrate in a large casserole (5½-inch). Boil to expel the hydrogen sulphide and then oxidize the iron by the cautious addition of a few cubic centimeters of strong nitric acid to the boiling liquid. Now add 5 cc. of strong hydrochloric acid (to provide for the formation of sufficient ammonium chloride to prevent the subsequent precipitation of alkaline earths as carbonates), then 15–20 cc. of strong bromine water (to remove manganese), and finally make alkaline with ammonia. Boil for a few moments, allow to settle somewhat, and then filter, washing with hot water. Reserve the filtrate. Continue from this point as described for calcium in X, 2 and X, 3, until the filtrate from the

calcium oxalate is obtained, or the combined filtrates, if two precipitations were made.

2. In my own laboratory, for ordinary technical work, especially when the amount of magnesium is supposed to be small, this filtrate is concentrated considerably by boiling in a large porcelain casserole and the magnesium is then precipitated by Handy's method (10). The precipitated magnesium ammonium phosphate is then ignited and weighed as $Mg_2P_2O_7$, as described in 4.

For more exact results proceed as follows:

3. Concentrate the filtrate from the calcium oxalate to small bulk by boiling in a large porcelain casserole, and when salts show a tendency to separate, transfer to a platinum dish or smaller porcelain casserole, in separate small portions if necessary, and evaporate on the water-bath to complete dryness. When dry, ignite gently to expel the ammonium salts. Cool, dissolve the residue in a very little dilute hydrochloric acid, make very faintly alkaline with ammonia and filter if necessary. The magnesium is now precipitated by the method of W. Gibbs.*†

4. Heat the solution to boiling and add, drop by drop, a solution of sodium ammonium phosphate ($NaNH_4HPO_4.2H_2O$), 160 grams to the liter,‡ until no further precipitate is produced. Most of the magnesium is at once precipitated as amorphous, dimagnesium phosphate ($MgHPO_4$). Allow the solution to cool, and then, stirring constantly, add about one-third its volume of ammonia. By this procedure the precipitate is changed to crystalline magnesium ammonium phosphate ($MgNH_4PO_4$) and the magnesium remaining in solution is precipitated in the same

---

* Am. Jour. Sci., (3) 5, 114.

† I usually prefer to precipitate by Handy's method, finally weighing the pyrophosphate, however, as described by Gibbs.

‡ 27 cc. of this solution will precipitate 0.5 gram of magnesium.

form. Let the cold mixture stand 2 or 3 hours and then decant the supernatant solution through a filter. Wash the precipitate three times by decantation with 2½ per cent. ammonia and then transfer to the filter and wash thoroughly with 2½ per cent. ammonia. Dry the filter and precipitate, transfer the latter as completely as possible to a weighed platinum crucible, burn the filter-paper in a platinum spiral and add the ash to the precipitate in the crucible. Cover the crucible and heat it, gently at first to expel the ammonia, and then over the blast-lamp until the residue is pure white. Cool in desiccator and weigh as $Mg_2P_2O_7$. This weight, multiplied by 0.2188 will give that of the Mg, or, if multiplied by 0.3624, that of the MgO.

5. In the above method it is not usually considered necessary to remove, as sulphide, any zinc present, inasmuch as the ignition to expel ammonium salts will tend to drive off the zinc as chloride. Should a little still remain, it is not likely to precipitate as phosphate on account of the presence of the ammonium and ammonium salts added.

6. **Note on the Precipitation of Magnesium Ammonium Phosphate.** — If the precipitation is made in a strongly ammoniacal solution instead of as described, some tribasic magnesium phosphate ($Mg_3P_2O_8$) will almost invariably form. This will be unchanged by the ignition and cause low results. It is, therefore, necessary to add the excess of ammonia later, but even in this case, if ammonium salts are present, the precipitate will always contain monomagnesium ammonium phosphate ($Mg(NH_4)_4(PO_4)_2$), which requires intense heating to constant weight to be certain it has all changed to $Mg_2P_2O_7$.

7. H. Neubauer,* whose experiments demonstrated the above facts, proceeds as follows: Slightly acidify the filtrate from

---

* Zeit. f. angew. Chem., 1896, 439. Treadwell, Quant. Anal., Hall., 2d Ed., p. 65.

the calcium oxalate with hydrochloric acid and then add an excess of sodium phosphate solution and one-third the solution's volume of 10 per cent. ammonia, stirring constantly. Allow to stand 24 hours and then wash by decantation, through a filter, with a little 2½ per cent. ammonia. It is better to evaporate the filtrate from the calcium oxalate to dryness and expel the ammonium salts by ignition, as previously described (3). If this is done, the solution may be filtered after only 4 hours' standing instead of 24. Place the beaker containing most of the precipitate under the funnel and pour through the filter a little dilute hydrochloric acid, washing the filter with hot water. Enough acid should be used to effect the solution of the precipitate in the beaker. Add now some ammonium chloride, a few drops of sodium phosphate solution, and one-third of the solution's volume of 10 per cent. ammonia. Allow to stand 4 hours. Pour the solution through a filter, wash the precipitate three times by decantation with 2½ per cent. ammonia, then transfer it to the filter and wash it thoroughly with the dilute ammonia. Dry and ignite the precipitate as previously described (4).

8. Limestones, Silicates, etc.—In material of these classes the members of the hydrogen sulphide group are usually absent, and, accordingly, no steps are necessary to remove them. When the substance is decomposable by acids the procedure is, with this exception, the same as previously described for ores (1). The treatment of silicates is begun as described for calcium on similar material (X, 7), but inasmuch as the magnesium is to be determined gravimetrically it is necessary to evaporate the acid solution to dryness, and remove the silica with the usual precautions (XXIV, 11). The acid solution, free from silica, is then treated as described for ores, usually omitting the hydrogen sulphide treatment.

9. **Handy's Volumetric Method.*** — This is a modification of Stolba's method in which the precipitated magnesium ammonium phosphate is titrated with standard sulphuric acid, the reaction being

$$MgNH_4PO_4 + H_2SO_4 = MgSO_4 + NH_4H_2PO_4.$$

A measured excess of sulphuric acid is used and titrated back with standard sodium hydroxide. The results obtained are very satisfactory.

The ore or other material is treated by the usual methods and the filtrate from the calcium oxalate precipitation is then treated as follows:

10. Add ammonia (sp. gr. 0.90) equivalent to 1/10 of the solution. Cool in water to 20° to 25° C. Precipitate by adding slowly with constant stirring a saturated solution of sodium ammonium phosphate, using 1 cc. for each 0.01 gram magnesium oxide. Stir vigorously for 5 minutes or shake in a flask for an equal length of time. In the former case let the solution stand until the clarification of the upper liquid shows that the reaction is complete. In the case of flask precipitations, if over 0.002 gram of magnesium oxide is present the solution may be filtered in 15 minutes. Suction may be used if desired, but if many solutions are to be filtered at once little is gained by its use. Use 10 per cent. ammonia wash (1 part ammonia (sp. gr. 0.90) to 9 of water). Deliver it preferably from an aspirator bottle placed about 4 feet above the bench. Wash by decantation as far as possible. Finally, wash the precipitate which has gone on the filter back into the beaker, stir it up with the ammonia wash and bring it again completely on the filter-paper. Wash once more, leaving the upper edge of the filter clear of precipitate so that it can be handled. Avoid assembling all of the precipitate

* James Otis Handy, Jour. Am. Chem. Soc., XXII, p. 31.

in the apex, but leave it fairly evenly distributed over the lower two-thirds of the paper. Allow the precipitates to drain and then transfer each in turn to a dry 5-inch filter-paper, allowing them to remain there open and face upward until the bulk of the moisture has been absorbed. After about 3 minutes transfer them to fresh dry filters, and in the case of heavy precipitates to a third set a few minutes later. Then place the filters on a shelf to dry at the temperature of the room, or place filter-papers and backing on the grating of an air-oven having a temperature of 50° to 60° C. After 15 or 20 minutes in the oven, or 45 minutes in the air, watch for the time when the filters have dried inward half an inch from the margin. This appearance has been found to indicate that the evaporation has gone far enough to insure the expulsion of the free ammonia. Now place the precipitates and filters in small dry beakers and treat each with a measured excess of decinormal sulphuric acid, stirring until the papers are disintegrated and the precipitates dissolved. Then add 2 drops of a 0.1 per cent. alcoholic solution of methyl orange. If this gives a clear decided pink, enough acid has been added. If it is only faintly pink, the excess of acid is slight and some minute particles may have escaped solution. In such cases add 5 cc. more decinormal sulphuric acid and stir well. Finally, dilute to about 100 cc. and titrate back with decinormal sodium hydroxide solution to the appearance of a clear yellow color, free from all suggestion of pink.

$$1 \text{ cc. } N/10\ H_2SO_4 = 0.002 \text{ gram } MgO.$$

If the filtrations have taken place during the latter part of the day, the filters may be removed from the funnels and allowed to stand overnight, after which they are titrated as described.

11. Notes on the above Method (Handy).—If the drying of the precipitate proceeds too far, solution in decinormal sulphuric acid is slow. If, however, the drying is stopped at the point described, there is enough moisture left so that, on stirring, the precipitate blends quickly with the acid and soon dissolves.

The tendency of magnesium to precipitate with iron and aluminum and with calcium oxalate must be met by re-solution in hydrochloric acid and reprecipitation. When the amount of calcium is considerable it is best to burn off the first oxalate precipitate before dissolving in hydrochloric acid. By this means the oxalate is decomposed and the addition of ammonia alone does not cause its sudden reprecipitation. Even in the second precipitation, if the boiling is allowed to proceed longer than is necessary to make the finely crystalline calcium oxalate settle well, some magnesium oxalate is sure to precipitate, betraying its presence by its coarser texture. The solution for magnesium precipitation usually does and always should contain in the form of ammonium chloride the equivalent of 5 cc. of con centrated hydrochloric acid per 100 cc.

CHAPTER XVIII.

MANGANESE.

1. THE following methods for the determination of manganese are applicable to most ores, the first one being ordinarily employed in my own laboratory. In unusual cases, where the material fails to be sufficiently decomposed by simple acid treatment, the mode of attack calls for the exercise of the operator's judgment. It is usual in such cases to employ one of the methods described under IRON (XV, **17, 18,** et seq.), and eventually to bring the solution of the substance into a proper condition for continuing by the regular method.

2. **Usual Method for Ores, etc.**—Treat 0.5 gram of the substance in an 8-oz. flask with whatever acids are necessary to decompose it. For an oxidized ore, about 10 cc. of strong hydrochloric acid are usually sufficient. With mixed ores it is best, in most cases, to start with hydrochloric acid, to dissolve the oxides, and then add 5–10 cc. of strong nitric acid to decompose the sulphides. With pure, or nearly pure, sulphides, begin at once with 10 cc. of nitric acid.

3. Heat very gently at first until the decomposition is complete. Finally add about 5 cc. of strong sulphuric acid and heat strongly, best over a free flame, until only a very little sulphuric acid remains. Cool and add about 100 cc. of water. Boil the mixture a moment and allow to stand, hot, with occa-

sional shaking, until anhydrous ferric sulphate, etc., has dissolved. Now add an excess of zinc oxide in thick emulsion.* Avoid a large excess, but add sufficient to precipitate all the iron, so that on standing a moment the mixture begins to settle clear and some zinc oxide can be seen in the bottom of the flask.† The mixture should be agitated to facilitate the precipitation, and it is best to boil it. Filter, wash thoroughly with hot water and receive the filtrate in a 500-cc. beaker.

4. Instead of precipitating and filtering as above, the following procedure is usually much quicker and quite accurate enough for technical purposes: Transfer the solution of the sulphates to a 200-cc. measuring-flask, using hot water. Now add the zinc oxide as described and then cool the mixture to room temperature under the tap. Make up to the mark with cold water, mix thoroughly and then filter off 100 cc. through a dry filter into a 100-cc. measuring-flask. Or, allow to settle sufficiently and remove 100 cc. with a pipette. Transfer the 100 cc. to a 500-cc. beaker.

5. To the solution in the large beaker add 3-4 grams of sodium acetate and about 25 cc. of saturated bromine water. Boil the mixture several minutes, adding more bromine water if there does not appear to be an excess. Allow the precipitate to settle somewhat and then filter and wash thoroughly with hot water. The filtrate should be clear and may be tested further, if desired, by adding more bromine water and boiling. A clear yellowish filtrate

---

* For ordinary use I simply mix the usual C. P. zinc oxide with water. A better and purer mixture is made as follows: Precipitate a solution of pure zinc sulphate with a solution of potassium hydroxide in insufficient amount to cause the solution to become alkaline. Wash the residue several times with hot water and then transfer it to a tightly stoppered bottle with enough water to hold it in suspension.

† If lead is present a little will remain in solution and be counted as manganese. It may be removed by adding 2 or 3 cc. of strong solution of potassium dichromate at this stage.

usually indicates that the manganese has been all precipitated, the yellow color showing an excess of bromine.

6. Place the washed precipitate, together with the filter, back in the beaker and run in from a burette what is judged to be an excess of the standard oxalic acid solution described below (9). On the basis of 0.5 gram of ore taken for assay, each cubic centimeter of this solution is equivalent to about 1 per cent. of manganese. Now add about 50 cc. of dilute (1:9) sulphuric acid and heat the mixture nearly to boiling with gentle agitation. Avoid shredding the filter. If the precipitate fails to dissolve completely, add more of the oxalic acid solution but avoid a large and unnecessary excess. When the precipitate has entirely dissolved, dilute the solution to 150–200 cc. with hot water and titrate to a faint permanent pink tinge with a standard solution of potassium permanganate (8). The excess of oxalic acid not consumed by the MnO_2 is thus found. Subtracting this from the total amount of oxalic acid used, the remainder is the amount used in reducing the MnO_2 to MnO. Multiply this figure by the value of 1 cc. in manganese to obtain the amount of manganese in the ore.

Example.—Took 0.5 gram of ore and did not divide the solution.

1 cc. of the permanganate solution = 0.5611 cc. of the oxalic acid sol.

1 cc. of the oxalic acid solution = 0.00502 gram of manganese, or, 1.04 per cent. when 0.5 gram of ore is taken.

Permanganate used in titrating excess of oxalic acid, 5.65 cc., equivalent to 3.17 cc. of oxalic acid solution.

Total number of cc. of oxalic acid solution used..	25.40
Less excess	3.17
Consumed by the MnO_2.	22.23

22.23 multiplied by 1.04 gives 23.12, the percentage of manganese in the ore. If the solution had been divided, the figure thus obtained would have to be multiplied by 2.

7. The reactions that take place may be expressed as follows:

Between the MnO_2 and the oxalic acid:

$$MnO_2 + C_2O_4H_2.2H_2O + H_2SO_4 = MnSO_4 + 2CO_2 + 4H_2O.$$

Thus 126.048 parts, by weight, of oxalic acid in the standard solution correspond to 55 parts of manganese. 1 per cent. of manganese, on the basis of 0.5 gram of ore taken for assay, is 0.005 gram, and the amount of oxalic acid corresponding to this is 0.011458 gram. This, then, is the amount to be contained in 1 cc., or, 11.458 grams per liter, in order that 1 cc. = 1 per cent. Mn.

Between the oxalic acid and permanganate:

$$2KMnO_4 + 5C_2O_4H_2.2H_2O + 3H_2SO_4$$
$$= K_2SO_4 + 2MnSO_4 + 10CO_2 + 10H_2O.$$

The standard solutions required are prepared as follows:

8. **Standard Potassium Permanganate.**—This may be approximately one-tenth normal, or about 3.16 grams per liter. The solution used for iron is ordinarily employed. The iron value multiplied by 1.128 will give the oxalic acid value. This value may also be determined by standardizing against oxalic acid or an oxalate directly. While the usual C. P. oxalic acid is not absolutely pure, it will generally suffice for technical work. Weigh carefully about 0.2 gram of the clean crystals and dissolve in a 6-oz. flask in a mixture of about 5 cc. of strong sulphuric acid and 100 cc. of water. Heat to 60°–70° C. and titrate with the permanganate solution to a permanent faint pink tinge. From the number of cubic centimeters used calculate the value of 1 cc. in oxalic acid.

Example.—Took 0.2195 gram of oxalic acid. Used 34.0 cc. of permanganate. Then $0.2195 \div 34.0 = 0.006455$, the value in grams of oxalic acid of 1 cc. of the permanganate solution. 1 cc. of an exactly decinormal permanganate solution would equal 0.0063024 gram of oxalic acid.

9. Standard Oxalic Acid Solution.—This, as explained above (7), should contain about 11.46 grams of $C_2O_4H_2.2H_2O$ per liter, in order that 1 cc. may equal about 1 per cent. of manganese when 0.5 gram of ore is taken for assay. Standardize as follows: Place in an 8-oz. flask about 5 cc. of strong sulphuric acid and dilute with 100 cc. of water. From a burette run in about 25 cc. of the oxalic acid solution. Heat the mixture to about 70° C. and titrate with the standard permanganate to a permanent faint pink tinge. From the number of cubic centimeters of permanganate used, and the known value of each cubic centimeter in oxalic acid, calculate the true strength of the oxalic acid solution and the consequent value of 1 cc. in manganese.

Example.—Ran in 21.70 cc. of oxalic acid solution. Used 40 cc. of permanganate. If the value of 1 cc. of permanganate is 0.006455 gram of $C_2O_4H_2.2H_2O$, then the 40 cc. used are equal to 0.2582 gram. As this is the amount of oxalic acid contained in 21.70 cc., each cubic centimeter contains 0.0119 gram. In the reaction between manganese dioxide and oxalic acid previously given (7), it is shown that 126.048 parts by weight of oxalic acid are equal to 55 parts of manganese. Consequently, to find the manganese value of the present oxalic acid, we have the proportion

$$126.048 : 55 = 0.0119 : x$$

$$x = 0.0052$$

$$(\text{Oxalic acid} \times 0.4364 = \text{Mn}).$$

1 cc. of the oxalic acid solution therefore equals 0.0052 gram of Mn, or, 1.04 per cent. on the basis of 0.5 gram of ore.

10. In determining the manganese value of the oxalic acid solution, the value of 1 cc. of the permanganate solution in cubic centimeters of standard oxalic acid solution should also be noted. Thus, in the last example, 40 cc. of permanganate solution were equal to 21.70 cc. of oxalic acid solution, therefore, 1 cc. of permanganate solution is equal to 0.5425 cc. of oxalic acid solution. This value should be marked on the permanganate bottle. The oxalic acid solution very slowly decomposes on keeping and requires to be restandardized occasionally. The addition of 50 cc. of sulphuric acid per liter greatly improves its keeping qualities.

10a. **Determination of MnO_2 only**—See Appendix.

11. **Volhard's Method.**—This is the method most generally used in western laboratories. Treat 1 gram of the ore in an 8-oz. flask with whatever acids are necessary to decompose it, beginning with 10 cc. of strong hydrochloric acid, and heating very gently, if oxides are present, and afterwards adding nitric acid, if necessary, to decompose sulphides and peroxidize iron. When the decomposition is complete add about 7 cc. of strong sulphuric acid and heat over a free flame until fumes of sulphuric anhydride are evolved copiously. Cool, add 25 cc. of water, boil a short time, and allow to stand, hot, with frequent shaking, until any anhydrous ferric sulphate has all dissolved. Transfer the mixture to a 500-cc. graduated flask and add an emulsion of zinc oxide* in slight excess to precipitate the iron. Agitate the flask to facilitate the precipitation and see that a slight excess of zinc oxide remains when the reaction is complete. Now dilute the contents of the flask up to the mark with cold water, mix thoroughly and allow to stand a short time and partially settle. By means of a graduated pipette draw off 100 cc. of the clear supernatant liquid and transfer it to an 8-oz. flask. While the precipitate in the

* See foot-note to 3.

500-cc. flask may appear large, it actually occupies but a very small space, and any error caused by it may consequently be neglected. Likewise the error in measurement due to change of temperature during the manipulation is insignificant. Heat the solution in the small flask to boiling, add two or three drops of nitric acid (which causes the subsequent precipitate to settle more quickly) and titrate with a standard solution of potassium permanganate. The permanganate causes a precipitate which obscures the liquid, and it is therefore necessary to titrate cautiously and agitate the flask after each addition, and then allow the precipitate to settle sufficiently to observe whether or not the solution is colored pink. A little experience will enable one to judge by the volume of the precipitate formed, about how rapidly to run in the permanganate. The final pink tinge, indicating the end of the reaction, is best observed by holding the flask against a white background and observing the upper edges of the liquid. When this point is attained, bring the contents of the flask nearly to a boil once more and again observe if the pink tint still persists, adding more permanganate if necessary. In making this end-test avoid actually boiling the liquid, as a continual destruction of the color may sometimes thus be effected and the true end-point considerably passed. When the color thus remains permanent the operation is ended. Observe the number of cubic centimeters of permanganate solution used and calculate the result.

12. The precipitate formed is not manganese dioxide, although the calculation can be correctly made as though the following reaction took place:

$$3MnSO_4+2KMnO_4+2H_2O=5MnO_2+K_2SO+2H_2SO_4.$$

In the presence of metallic salts, such as those of calcium or zinc, manganites of varying composition are formed, e.g.,

$$4KMnO_4+5ZnSO_4+6MnSO_4+14H_2O$$
$$=4KHSO_4+7H_2SO+5ZnH_2.2MnO_3.$$

The precipitated manganese is, however, always in the tetravalent form, and, therefore, the ratio of the first reaction between the manganese and the permanganate is not changed.

13. It is customary to use the same permanganate solution for both iron and manganese. Having determined the factor for iron (xv, 12), this may be multiplied by 0.2952 to obtain the factor for manganese. It will be observed in the first equation above that $2KMnO_4$ are required for 3Mn, and in the reaction for iron in xv, 2, that $2KMnO_4$ are required for 10Fe. Therefore 559 parts of iron are equivalent to 165 parts of manganese, or, 1 part of iron to 0.2952 part of manganese.

14. The Bismuthate Method for the Determination of Manganese.*—This method is based on the fact that a manganous salt in the presence of an excess of nitric acid is oxidized to permanganic acid by bismuth tetroxide. The permanganic acid formed is very stable in nitric acid of 1.135 sp. gr. when the solution is cold, but in hot solutions the excess of the bismuth tetroxide is rapidly decomposed and then the nitric acid reacts with the permanganic acid, and as soon as a small amount of manganous salt is formed the remainder of the permanganic acid is decomposed, manganous nitrate dissolves and manganese dioxide precipitates.

In the cold, however, the excess of the bismuth salt may be filtered off and to the clear filtrate an excess of ferrous sulphate

* Andrew A. Blair, Jour. Am. Chem. Soc., XXVI, 793. Method originally proposed by Schneider and modified first by Reddrop and Ramage and then by Brearley and Ibbotson.

added and the amount necessary to deoxidize the permanganic acid determined by titrating with permanganate. The end reactions are very sharp and the method is extremely accurate, but the presence of even a trace of hydrochloric acid utterly vitiates the results. As pointed out by Reddrop and Ramage, bismuth tetroxide, which was used by Schneider, is difficult to obtain free from chlorides and they recommended sodium bismuthate, which they prepare as follows: Heat 20 parts of caustic soda nearly to redness in an iron or nickel crucible and add, in small quantities at a time, 10 parts of basic bismuth nitrate, previously dried in a water-oven. Then add 2 parts of sodium peroxide and pour the brownish yellow fused mass on an iron plate to cool; when cold, break it up in a mortar, extract with water and collect on an asbestos filter. The residue, after being washed four or five times by decantation, is dried in the water-oven, then broken up and passed through a fine sieve. (The Baker & Adamson Chemical Co. have prepared sodium bismuthate in this manner which is perfectly free from manganese chlorides and has proved entirely satisfactory.)

The Method.

Steels. — Dissolve 1 gram of drillings in 50 cc. of nitric acid (sp. gr. 1.135) in an Erlenmeyer flask of 200 cc. capacity. Cool and add about 0.5 gram of bismuthate. The bismuthate may be measured in a small spoon, and experience will soon enable the operator to judge of the amount with sufficient accuracy. Heat for a few minutes, or until the pink color has disappeared, with or without the precipitation of manganese dioxide. Add sulphurous acid, solution of ferrous sulphate or sodium thiosulphate in sufficient amount to clear the solution and heat until all nitrous oxide has been driven off. Cool to about 15° C., add an excess of bismuthate and agitate for a few minutes.

Add 50 cc. of water containing 30 cc. of nitric acid to the liter and filter through an asbestos felt on a platinum cone into a 300-cc. Erlenmeyer flask and wash with 50 to 100 cc. of the same acid. The arrangement shown in Fig. 17 has proved very satisfactory. Run into the flask from the pipette, shown in Fig. 18,

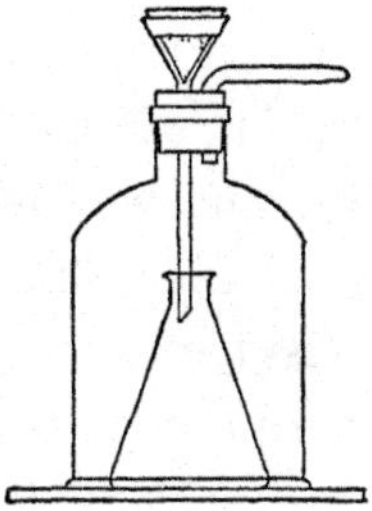

FIG. 17.

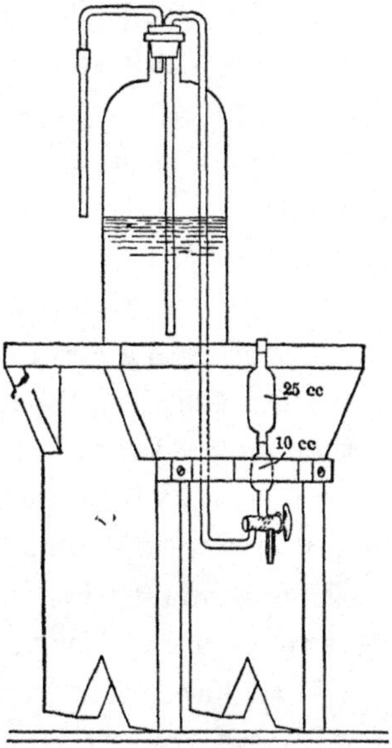

FIG. 18.

a measured volume of ferrous sulphate solution and titrate to a faint pink color with permanganate. The number of cubic centimeters of the permanganate solution obtained, subtracted from the number corresponding to the volume of ferrous sulphate solution used, will give the volume of permanganate equivalent to the manganese in the sample, which, multiplied by the value of the permanganate in manganese, gives the amount of manganese in the steel.

Pig Iron. — Dissolve 1 gram in 25 cc. of nitric acid (sp. gr. 1.135) in a small beaker and as soon as the action has ceased, filter on a 7 cm. filter into a 200 cc. Erlenmeyer flask, wash with 30 cc. of the same acid and proceed as in the case of steels.

In the analysis of white irons it may be necessary to treat the solution several times with bismuthate to destroy the combined carbon. The solution, when cold, should be nearly colorless; if not, another treatment with bismuthate is necessary.

Iron Ores Containing Less than 2 Per Cent. of Manganese. — Treat 1 gram in a platinum dish or crucible with 4 cc. of strong sulphuric acid, 10 cc. of water and 10 to 20 cc. of hydrofluoric acid. Evaporate until the sulphuric acid fumes freely. Cool and dissolve in 25 cc. of 1.35 nitric acid. If no appreciable residue remains, transfer to a 200 cc. Erlenmeyer flask, using 25 cc. of 1.135 nitric acid to rinse the dish or crucible and proceed as usual. If there is an appreciable residue, filter on a small filter into a beaker, wash with water, burn the filter and residue in a crucible and fuse with a small amount of potassium acid sulphate. Dissolve in water with the addition of a little nitric acid, add to the main filtrate, evaporate nearly to dryness, take up in 1.135 nitric acid and transfer to the flask as before.

Manganese Ores and Iron Ores High in Manganese. — Treat 1 gram as in the case of iron ores, using a little sulphurous acid if necessary. Transfer the solution to a 500 cc. flask, dilute to the mark, mix thoroughly and measure into a flask from a carefully calibrated pipette such a volume of the solution as will give from 1 to 2 per cent. of manganese and enough strong nitric acid (sp. gr. 1.4) to yield a mixture of 1.135 acid in a volume of 50 to 60 cc. For example, in a 50 per cent. ore use 10 cc. of the solution and add 30 cc. of water and 10 cc. nitric acid (sp. gr. 1.4). In this case the manganese must be calculated on $\frac{1}{50}$ of a gram or 20 mg. of ore. When working on such amounts it is always desirable to make duplicate analyses and take the mean, as a difference of 0.1 cc. makes a large error in the result. When the ore contains a much smaller amount of manganese, say 5 to 10 per cent., it is better to make up the solution to say 100 cc. instead of 500.

Ferro-manganese. — Treat 1 gram exactly like steel. Dilute to 500 or 1000 cc. and proceed as in manganese ores.

Ferro-silicon. — Treat 1 gram with sulphuric and hydrofluoric acids and proceed as with iron ores.

Special Steels. — Steels containing chromium offer no special difficulties except that it must be noted that while in hot solutions the chromium is oxidized to chromic acid, which is reduced by the addition of sulphurous acid, the oxidation proceeds so slowly in cold solutions that if there is no delay in the filtration and titration the results are not affected. Steels containing tungsten are sometimes troublesome on account of the necessity for getting rid of the tungstic acid. Those that decompose readily in nitric acid may be filtered and the filtrate treated like pig iron, but when it is necessary to use hydrochloric acid it is best to treat with *aqua regia*, evaporate to dryness, redissolve in hydrochloric acid, add a few drops of nitric acid, dilute, boil, and filter. Get rid of every trace of hydrochloric acid by repeated evaporations with nitric acid and proceed as with an ordinary steel.

Reagents.

Nitric Acid (sp. gr. 1.135). — A mixture of 3 parts of water and 1 part strong nitric acid answers perfectly for this purpose.

Nitric Acid (3 *per cent.*). — Thirty cc. of strong nitric acid to the liter.

Permanganate Solution and Ferrous Sulphate Solution.—One gram of potassium permangate to the liter gives a solution of convenient strength, and 12.4 grams of ferrous ammonium sulphate and 50 cc. of strong sulphuric acid,* made up to 1 liter, gives a solution which is almost exactly equal to the permanganate solution. As the strength of the ferrous sulphate solution changes quite rapidly, while the permanganate remains unaltered for months, it is unnecessary and troublesome to attempt to keep them of the same strength. By using a constant volume of the ferrous sulphate solution and testing it against the permanganate solution every day, the calculation of the results is very simple. It is necessary that the conditions should be the same in getting the strength of the ferrous sulphate solution as in titrating a solution for manganese, and after many experiments the following method of procedure was adopted: Measure into a 200 cc. flask 50 cc. of nitric acid (sp. gr. 1.135), cool and add a very small amount of bismuthate, dilute with 50 cc. of 3 per cent. nitric acid, filter into a 300 cc. flask and wash with 50 cc. of 3 per cent. nitric acid. If the felt is well coated with bismuthate it is unnecessary to add any to the nitric acid in the flask, as filtration through the mass of bismuthate on the felt will answer the purpose. Run in from the pipette (Fig. 18) 25 cc. of ferrous sulphate solution and titrate with the permanganate to a faint pink. This gives the value in permanganate of the ferrous sulphate solution. With this method of procedure the discrepancies that had occurred entirely disappeared, and it is possible to make any number of determinations with a variation of less than 0.1 cc.

The permanganate solution may be standardized in three ways:

First, by getting its value in iron in the usual way and calculating its value in manganese. The proportion is 279.5 : 55, or as 1 : 0.1968.

* Dr. C. B. Dudley proposes to use 25 cc. of sulphuric and 25 cc. of strong phosphoric acid as tending to give a more nearly colorless solution.

Second, by titrating a steel with a known amount of manganese and getting the value of the solution by dividing the percentage of manganese by the number of cubic centimeters of the permanganate used.

Third, by making a solution of pure manganese sulphate and determining the manganese in it by evaporating a weighed amount of the solution to dryness, heating to dull redness and weighing as manganese sulphate, which, multiplied by 0.36409 gives the amount of manganese. 5 grams of "C. P." manganese sulphate dissolved in 500 cc. of water and filtered will give a solution containing about 0.0035 gram of manganese to the gram of solution. Weigh 1 to 3 grams of the solution in a crucible, transfer to a 200 cc. flask, using 50 cc. of nitric acid (sp. gr. 1.135), cool, add 0.5 to 1 gram bismuthate and allow it to stand for 3 or 4 minutes, shaking at intervals. Add 50 cc. of 3 per cent. nitric acid and filter through the asbestos filter and wash with 50 or 60 cc. of the same acid. Run 25 cc. of the ferrous sulphate solution into the flask from the pipette and titrate with the permanganate solution to a faint pink. Subtract the number of cubic centimeters of the permanganate solution obtained from the value of the 25 cc. of ferrous sulphate solution in permanganate and the result is the number of cubic centimeters of the permanganate corresponding to the manganese in the manganese sulphate solution used. Divide the weight of the manganese in the manganese sulphate used by the number of cubic centimeters of permanganate, and the result is the value of 1 cc. of permanganate in manganese.

Example. — One gram manganese sulphate solution contains 0.003562 gram manganese; 2.0372 grams manganese sulphate solution equals 0.0072565 gram manganese; 25 cc. ferrous sulphate solution equals 24.5 cc. permanganate solution; 2.0372 grams manganese sulphate, after oxidation and addition of 25 cc. ferrous sulphate solution, require 3.6 cc. permanganate solution; 24.5 cc. − 3.6 cc. = 20.9 cc.; 0.0072565 divided by 20.9 = 0.0003472, or 1 cc. of permanganate equals 0.0203472 gram of manganese. If then 1 gram of steel, after oxidation and addition of 25 cc. ferrous sulphate solution, requires 6.2 cc. permanganate solution to give the pink color, 24.5 − 6.2 = 18.3 × 0.0003472 = 0.006354 gram, or the sample contains 0.635 per cent. manganese.

Notes and Precautions.

The delicacy of the reaction of manganese in nitric acid solution with sodium bismuthate is extraordinary. 0.000005 gram of manganese gave an appreciable color in 50 cc. of solution.

When the proper precautions are observed, this method for materials containing small amounts of manganese, say up to 2 per cent., is more accurate than any other method, volumetric or gravimetric, that I have ever used.

As will be seen in the description of the various methods of solution, the use of hydrochloric acid has been avoided because the presence of even traces of this reagent is fatal to the accuracy of the method. Where it is impossible to avoid its use and its presence is suspected in the final nitric acid solution, the addition of a drop or two of silver nitrate will overcome the difficulty, but the filter must be rejected after using it for filtering a solution so treated.

Any form of asbestos filtering tube may be used for filtering off the bismuthate, but the perforated cone with bell jar, shown in Fig. 17, is the most satisfactory, because it has the largest area of filtering surface. One filter may be used for fifty or more determinations and the time occupied in filtering and washing one determination is only from one minute and a half to three minutes. The filtrate must be perfectly clear, for the least particle of bismuthate carried through will vitiate the result by reacting with the excess of ferrous sulphate. As soon as the filtration and washing are completed, the ferrous sulphate should be added and the excess titrated with the permanganate solution, as the permanganic acid gradually decomposes on standing and the warmer the solution the more rapid is the decomposition. At a temperature of 5° C. the solution will remain unaltered for several hours, but at 40° C. fifteen minutes will show an appreciable change. The larger the amount of manganese the more rapid the change. It is especially important not to allow the solution to stand after adding the ferrous sulphate, as the excess of this reagent reacts with the nitric acid in a few minutes and the formation of the smallest amount of nitrous oxide is fatal to the accuracy of the determination. For this reason it is important to boil off every trace of nitrous oxide when in the earlier part of the operation sulphurous acid or other deoxidizing agent is added.

When working with steels of unknown manganese content, it may often happen that 25 cc. of ferrous sulphate solution are insufficient to entirely reduce the permanganic acid, in which case an additional amount of ferrous sulphate must be added and the pipette shown in Fig. 18 has been arranged to meet this contingency. It will be noticed that the solution of permanganic acid upon the addition of an insufficient amount of ferrous sulphate does not necessarily retain its pink or purple color, but usually changes to a dirty brown. When this occurs, the lower part of the pipette may be emptied directly into the flask and the value of the two parts taken as the amount from which the number of cubic centimeters of permanganate corresponding to the excess of ferrous sulphate must be subtracted. When the sample is low in manganese, the 10 cc. portion of the pipette alone may be used, so that the arrangement allows a great deal of variation in the manganese content of the samples worked on.

There is no advantage in using permanganate solutions differing in strength from the one given above, but the strength of the ferrous sulphate solution may be changed to meet special cases.

CHAPTER XIX.

MERCURY.

1. **Wet Method.**—*Applicable to cinnabar ores not containing appreciable amounts of other reducible metallic compounds.*

Treat 1–5 grams of the ore, according to its supposed richness, in a covered porcelain dish with sufficient *aqua regia* to decompose it. Heat gently until decomposition is complete, and then remove and wash the cover and evaporate the solution to dryness on a water-bath. Take up the residue in strong hydrochloric acid and again evaporate to dryness to expel all of the nitric acid. Again dissolve in hydrochloric acid, dilute sufficiently and filter. To the filtrate add an excess of a clear stannous chloride solution containing an excess of acid (which may be made by dissolving tin in an excess of hydrochloric acid), and boil the mixture for a short time. The mercuric chloride is reduced to finely divided metallic mercury. Allow the precipitate to settle completely and then decant the clear liquid carefully. Unite the precipitate into one globule by heating it with a little moderately dilute hydrochloric acid mixed with a few drops of stannous chloride. Wash the mercury by decantation, first with water slightly acidified with hydrochloric acid, and then with pure water, and then transfer it to a weighed porcelain crucible. Absorb as much of the adhering water as possible with filter-paper and then place the crucible in a desiccator over sulphuric acid. When perfectly dry weigh as metallic mercury.

The results obtained by this method are certain to be somewhat low owing to the impossibility of evaporating the hydrochloric acid solutions to dryness, in driving off the nitric acid, without volatilization of some mercuric chloride with the escaping steam. As a technical method, however, for low-grade cinnabar ores it answers very well.

2. **Dry Method.** *Applicable to all ores.*—Prepare a combustion-tube about 18 or 20 inches long and sealed at one end. Place in the closed end a column of crushed magnesite about 4 inches long, then add a section of about 2 inches of freshly ignited caustic lime, then about 4 inches of an intimate mixture of a weighed portion of the ore with an excess of caustic lime, rinsing out the mortar with a little lime and adding this also, and then about 2 inches of caustic lime. Finally, insert against the whole a loose plug of asbestos about 2 inches long. Now draw out the end of the tube to a narrow opening and bend it at right angles. Tap the tube horizontally on the table, gently, so as to leave a free passage for gases throughout its length. Place the tube in a combustion furnace and arrange a small flask party filled with water so that the point of the tube just touches the water, which thus closes it. Now proceed to heat the tube, beginning at the end containing the asbestos and gradually approaching the other end, until the tube is red-hot throughout the entire portion in the furnace. The carbon dioxide evolved from the magnesite serves to sweep out the last traces of mercury vapor into the water. While the tube is still hot cut off the point above the flask and rinse any condensed mercury into the water. Agitate the flask so as to collect the mercury into one globule, and, after allowing to stand and settle some time, decant off the perfectly clear water and transfer the mercury to a weighed

porcelain crucible. Remove as much of the adhering water as possible with filter-paper and then dry the mercury to constant weight over sulphuric acid in a desiccator. It must not be heated.

3. **Eschka's Method.*** *Particularly applicable to low-grade ores.*—Mix from 0.2 to 2 grams of the ore with from 1 to 4 grams of iron filings in a porcelain crucible of suitable size. Prepare a dish-shaped cover of thin sheet gold for the crucible, that may be kept cool by being filled with water. To prevent any mercury from escaping, the cover should be large enough to project somewhat over the edge of the crucible. Place the crucible in a ring-stand, put on the weighed cover and nearly fill it with water. Carefully heat the lower part of the crucible with a Bunsen burner, but keep the upper part cool. A ring of thin asbestos board fitted around the crucible at the proper height will assist materially in preventing the top from getting overheated. Add cold water to the cover from time to time. It will require from 10 to 30 minutes to distill off the mercury. When the operation is considered ended remove the gold cover, dry carefully without heating and weigh. The increase over the original weight represents the mercury. A silver cover appears to answer practically as well as one of gold

4. **Krieckhaus' Volumetric Method.**†—Weigh 2 grams of the ore into a 150-cc. beaker. Add 2 cc. of strong nitric acid and 10 cc. of strong hydrochloric acid. Measure the nitric acid first and then measure the hydrochloric without washing the glass. This ensures getting all of the nitric acid, which is essential, although an excess must be avoided.

Allow to stand cold for an hour or more, when the mercury

---

* Zeit. f. anal. Chem., Vol. 11, p. 344.

† Kindly furnished by Mr. Leon L. Krieckhaus. The description was received too late for experimental tests in my laboratory, but the results sent by Mr. Krieckhaus check closely with those he obtained by Eschka's method.

will all be in solution; or, to save time, heat slightly but not to boiling. Now dilute with 10–15 cc. of water, heat nearly to boiling and filter into a 300-cc. Erlenmeyer flask, washing with hot water. To the filtrate, which contains the mercury (and need not be much over 100 cc. in volume), add 60 cc. of stannous chloride solution (prepared as described below), cork the flask and allow to stand, tilted at an angle of about 45°, until the mercury has all settled and the liquid is clear. The settling takes about two hours.

Decant the liquid and then fill the flask about two-thirds full with cold water, slightly acidulated with sulphuric acid to prevent the separation of basic tin salts. Again allow to settle with the flask tilted as before. If any mercury floats, a jet of water will cause it to settle. In about 10 minutes decant the clear liquid as completely as possible. Dissolve the precipitated mercury in 2 or 3 cc. of strong nitric acid, warming gently to ensure complete solution and oxidation to mercuric nitrate.

Add about 75 cc. of cold water and a few cubic centimeters of a solution of ferric nitrate, as indicator (or ferric ammonium alum slightly acidulated with nitric acid), and titrate cold with standard potassium thiocyanate solution to the usual red tinge.

5. If the potassium thiocyanate solution contains 9.722 grams of the salt per liter, 1 cc. will equal 0.01 gram of mercury, or 0.5 per cent. on the basis of 2 grams of ore taken for assay. It is best to standardize it with pure silver or silver nitrate, as described in VI, 7. The silver value of 1 cc. multiplied by 0.9265 will give the mercury value.

Prepare the stannous chloride solution by mixing 50 grams of the salt, 50 cc. of strong hydrochloric acid and 150 cc. of water, and boiling with a stick of metallic tin until clear. Keep a stick of tin in the bottle.

6. Notes by Mr. Krieckhaus.—If lead is present in an ore, it can probably be largely removed by adding a little sulphuric acid before filtering off the gangue. If any remains in solution and comes down as chloride with the mercury, the chlorine it carries will interfere, but the lead chloride could probably be washed out of the mercury with hot water. Stannous chloride precipitates copper as cuprous chloride, but this may be easily removed with dilute ammonia.

In order that the mercury shall settle well it is best to have the solution cold and the tin solution quite strongly acid, but not saturated with stannous chloride. Add only 1 or 2 cc. of the tin solution at first and agitate the flask. This will cause a precipitation of mercurous chloride. Then add the remainder of the tin solution and allow to stand as above described. Under these conditions the mercury will settle clear in an hour or two. If the liquid is hot or the tin solution too strong, the mercury is so finely divided that it is slow to settle and difficult to wash without loss.

7. **Seamon's Volumetric Method.***—Weigh 0.5 gram of the finely ground ore into an Erlenmeyer flask of 125 cc. capacity. Add 5 cc. of strong hydrochloric acid and allow it to act for about ten minutes at a temperature of about 40° C., then add 3 cc. of strong nitric acid and allow the action to continue for about ten minutes longer. The mercury should now all be in solution. Now if lead be present, add 5 cc. of strong sulphuric acid; it may be omitted otherwise. Dilute with 15 cc. of water and then add ammonia cautiously until the liquid is slightly alkaline. Bismuth, if present, will precipitate. Acidify faintly with nitric acid, filter, receiving the filtrate in a beaker, and wash thoroughly.

Add to the filtrate 1 cc. of strong nitric acid that has been made brownish in color by exposure to the light, and titrate

---

* Manual for Assayers and Chemists, p. 112.

with a standard solution of potassium iodide until a drop of the liquid brought into contact with a drop of starch liquor, on a spot-plate, shows a faint bluish tinge. It is a good plan to set aside about one-third of the mercury solution and add it in portions as the end-point is successively passed, finally rinsing in the last portion and titrating to the end-point very carefully.

Deduct 0.5 cc. from the burette reading and multiply the remaining cubic centimeters used by the percentage value of 1 cc. in mercury to obtain the percentage in the ore.

The standard potassium iodide solution should contain 8.3 grams of the salt per liter. Standardize against pure mercuric chloride. Dissolve a weighed amount of the salt in water, add 2 cc. of the discolored nitric acid and titrate as above. 1 cc. of the standard solution will be found equivalent to about 0.005 gram of mercury, or about 1 per cent. on the basis of 0.5 gram of ore taken for assay.

The precipitate of red mercuric iodide which forms during the titration may not appear if the amount of mercury present is very small, but this failure to precipitate does not appear to affect the result.

Iron, copper, bismuth, antimony and arsenic, when added separately to the ore, did not influence the results in Seamon's tests. Silver interferes. Duplicate results should check within 0.1–0.2 of one per cent.

CHAPTER XX.

MOLYBDENUM.

(See Appendix.)

1. **Method for Ores.**—Fuse 1 gram of the ore in a platinum dish or crucible with 4 grams of sodium carbonate and 0.5 gram of potassium nitrate. Cool and extract the melt with hot water. If the solution shows any color due to manganese add a little alcohol and warm so as to effect its reduction and precipitation. Filter, washing with hot water. Nearly, but not quite, neutralize the filtrate with nitric acid. The amount to use should be ascertained by a blank test on exactly 4 grams of sodium carbonate. Evaporate the solution to approximate dryness, dilute, filter and add to the cold alkaline solution a solution of mercurous nitrate until it ceases to effect a further precipitation. The precipitate consists of mercurous molybdate, besides mercurous carbonate produced by the slight excess of alkali carbonate left in the solution. Any chromium, vanadium, tungsten, phosphorus, or arsenic present will also be precipitated. The mercurous carbonate serves to neutralize the free nitric acid always present in the mercurous nitrate. If the alkalinity was too great, as shown by the precipitate becoming unduly large and still increasing as mercurous nitrate is added, add nitric acid drop by drop until an added drop of mercurous nitrate produces no cloud. Heat the liquid to boiling and filter off the black precipitate. Wash with a dilute solution of mercurous nitrate. Dry the filter and pre-

cipitate and transfer the latter as completely as possible to a watch-glass. Dissolve the precipitate still remaining on the filter in hot dilute nitric acid and receive the solution in a large weighed porcelain crucible. Evaporate the solution to dryness on a water-bath and then add the main portion of the precipitate. Heat, without melting, over a very low flame until the mercury is completely volatilized. Cool and weigh as MoO_3. Multiply this weight by 0.6667 to obtain the weight of the molybdenum.

2. In case the ore contained chromium, vanadium, tungsten, phosphorus, or arsenic in appreciable amount, the ignited MoO_3 must be further treated as follows:

To avoid transference the ignition should be made in a small platinum dish. Fuse with a very little sodium carbonate and leach the residue with hot water and filter. To the filtrate add sulphuric acid in slight excess and transfer to a small pressure flask. Pass in hydrogen sulphide to saturation in the cold, then close the flask, heat on the water-bath until the precipitate has completely settled, allow to cool and filter. Wash first with very dilute sulphuric acid and then with alcohol until the acid is all removed. The precipitate consists of molybdenum and arsenic sulphides and sometimes contains also traces of platinum from the dish. Place the moist precipitate and filter in a large weighed porcelain crucible, dry thoroughly on a water-bath, then cover the crucible and heat very carefully over a small flame until volatile hydrocarbons are expelled. Now remove the cover and burn the carbon from the sides of the crucible at as low a temperature as possible. Finally, increase the heat gradually until the molybdenum sulphide is all changed to oxide, i.e., until no more sulphur dioxide is evolved. Cool, add a little mercuric oxide suspended in water, stir the mixture well, evaporate to dryness on the water-bath and then expel the mercuric oxide by gentle ignition and weigh the residue, after cooling, as MoO_3.

The mercuric oxide effects the oxidation of any unburned particles of carbon from the filter-paper.

3. If the color of the sulphide precipitate showed arsenic to be absent the last result may be accepted as correct. When arsenic is present, or in case of doubt, fuse the last residue in the crucible with a little mixed sodium and potassium carbonates at as low a temperature as possible and leach the melt with hot water. Filter, add one-third its volume of strong ammonia to the filtrate and then about 20 cc. of "magnesia mixture" (p. 213). Add the latter drop by drop with constant stirring. Allow to stand, covered, 12 hours in the cold and then filter. Wash well with 2½ per cent. ammonia. Acidify the filtrate with dilute sulphuric acid, transfer to a pressure flask and repeat the operations described in 2.

4. The presence of molybdenum may be shown in the final residue as follows: Heat the residue, or a little of it, in porcelain, with a single drop of strong sulphuric acid until the latter is nearly volatilized. On cooling, a beautiful blue color is proof of the presence of molybdenum.

CHAPTER XXI.

NICKEL AND COBALT.

NICKEL and cobalt require the same procedure at the outset, and when a separate determination of either or both is required it is made after the elements or their compounds have been separated together from interfering substances.

1. **Method for Ores, etc.**—To 0.5 gram of the ore in an 8-oz. flask add about 10 cc. of strong nitric acid and boil until the red fumes have about ceased coming off, or until the acid is about half gone. Then add about 5 grams of fine crystallized potassium chlorate and 5 cc. more nitric acid. Boil to complete dryness, but avoid overheating the residue. It is best to manipulate the flask over a free flame to save time and avoid loss by bumping. Towards the last, spread the pasty mass around on the inside of the flask so as to form a thin layer. When completely dry, cool and add 35 cc. of strong ammonia water and warm gently until the residue is disintegrated as thoroughly as possible. Now add about 15 cc. of strong bromine water and heat a while longer to precipitate any manganese in solution. Filter, washing with hot water. Rinse the residue on the filter back into the flask again, place the latter under the funnel and pour a little hot, dilute (1:2) hydrochloric acid through the filter to dissolve all remaining residue and wash with water. Boil the contents of the flask to completely dissolve the soluble portion of the residue,

dilute somewhat, add 15 cc. of strong bromine water and then ammonia in excess. Boil for a short time and then filter and wash with hot water thoroughly.

2. Unite the two filtrates, which now contain practically all the nickel and cobalt,* in a large casserole, and boil off most of the excess of ammonia. Make just acid with hydrochloric acid and then add 5 cc. in excess. Again heat to boiling and boil for a few minutes to decompose chlorates, etc. Dilute, if necessary, to about 250 cc. with hot water and pass in hydrogen sulphide until cold. Filter off the sulphides of copper, lead, etc., and wash with dilute hydrogen sulphide water. Receive the filtrate in a large casserole (5- or 6-inch.)

3. Boil the solution until the hydrogen sulphide is completely expelled and then add a solution of pure potassium hydroxide in excess and keep at nearly boiling temperature for some time. The color of the precipitate thus obtained will indicate the presence or absence of cobalt. If of a pure apple-green color, no cobalt, or only an insignificant amount is present. The cobalt precipitate is at first blue but soon changes to black, by contact of the liquid with the air, and thus the comparative amount of cobalt present can be judged by the degree of discoloration. Filter the precipitate and wash with hot water.

4. If the precipitate is of a pure apple-green color and very small in amount, with no evidence of the presence of alkaline-earthy carbonates, it will often suffice, after washing thoroughly, to ignite and weigh it as NiO. Multiply by 0.7858 to obtain the weight of the nickel. If simply discolored with cobaltic hydroxide, but otherwise apparently pure and small in amount,

---

* If considered desirable, the iron precipitate may be again redissolved and a basic acetate separation made, the filtrate from this being sufficiently evaporated and added to the other filtrates. In technical work I have rarely found this necessary.

it may be sufficient to ignite and weigh as NiO and report, ofter calculation, as combined nickel and cobalt.

5. If the amount of precipitate is large and apparently free from cobalt, it may still contain zinc, if the latter was present in the original substance, and alkaline-earth carbonates, precipitated by the carbonate in the caustic alkali used. In such a case it may sometimes suffice to redissolve and make a second precipitation as follows: Dissolve the precipitate on the filter in warm dilute (1:3) hydrochloric acid and receive the filtrate in the large casserole previously used. Dilute the solution sufficiently, heat to boiling, and add an excess of pure potassium hydroxide and some bromine water. The nickel is precipitated as brownish-black nickelic hydroxide, $Ni(OH)_3$. When the operation is complete filter off the precipitate, washing with hot water, first by decantation and then on the filter. Dry and ignite the residue and weigh as NiO. If the impurities present in the first precipitate were not large in amount they will be mostly eliminated by the second precipitation. The ignited NiO will contain no alkali, but usually has traces of silica.

6. If it is desired to determine the amount of silica mixed with the NiO, dissolve the latter in the crucible in hydrochloric acid, evaporate to complete dryness, moisten with strong hydrochloric acid and then take up with hot water. Filter through a small filter and wash with hot water. Ignite the moist filter and residue and weigh as SiO_2.

The above are given simply as approximate methods that will frequently serve where the ore is fairly free from interfering impurities and where any cobalt present may be reported as nickel.

7. For more exact results, including the separation of nickel and cobalt, proceed as follows: Begin as before (1) and continue as described, omitting, however, the use of bromine,

until the filtrate from the hydrogen sulphide precipitate is obtained. Boil this filtrate until the hydrogen sulphide is completely expelled and then add ammonia in slight excess. Now acidify slightly with acetic acid, add 1 or 2 grams of ammonium acetate, heat to 70°–80° C. and saturate with hydrogen sulphide. The nickel and cobalt are precipitated as sulphides. Filter, washing with hot water. The filtrate may still contain small amounts of nickel and cobalt. Concentrate it and add some colorless ammonium sulphide. Make slightly acid with acetic acid, warm and filter. If a precipitate is obtained, collect it on a separate filter. Repeat this testing of the filtrate until no further precipitation is produced.

8. Wash the precipitated sulphides from the filters, as completely as possible, into a small porcelain dish or casserole. Dry and burn the filters and add the ash also. Dissolve the whole in hydrochloric acid and a little nitric acid. The solution now contains nickel, cobalt, and possibly zinc. To remove the latter, evaporate to small volume, add 2 or 3 grams of pure finely crystallized ammonium chloride,* evaporate to dryness on a water-bath and then heat carefully until all the ammonium chloride is expelled. The zinc is driven off at the same time. When cool, dissolve the residue in nitrohydrochloric acid and expel the excess of acid by evaporating nearly to dryness. Dilute sufficiently, heat to boiling, add an excess of pure potassium hydroxide and some bromine water. Filter and wash thoroughly with hot water, first by decantation, then on the filter. Dry and ignite. If only nickel is present, or if nickel and cobalt are to be determined together, the residue may be weighed as NiO and the metal reported, after calculation (factor, 0.7858) as Ni, or, Ni+Co. Determine and deduct silica if desired (6).

* Fresenius. Zeit. für. Anal. Chem. Schæffer, Am. Chem., IV, 289. Fresenius prescribes 5 grams ammonium chloride for every 0.2 gram ZnO.

9. When the metals are to be determined separately,* precipitate the nickel and cobalt as hydroxides from the last solution above, without the addition of bromine. Heat nearly to boiling for some time and then filter and wash only once or twice with hot water. The two metals may now be separated by Liebig's mercuric oxide method, as modified by Wöhler. Wash the hydroxides from the filter into a large casserole, then place the latter under the funnel and pour a saturated solution of potassium cyanide over the filter to dissolve whatever precipitate still remains. Wash the filter with hot water and warm the filtrate and mixed hydroxides until solution is complete, adding more potassium cyanide if required but avoiding an unnecessary excess. Heat the solution for at least one hour on the water-bath, to convert the cobalt compound into potassium cobalticyanide. Now add to the hot solution an excess of finely pulverized red mercuric oxide and boil the mixture for an hour. The nickel is precipitated as nickelous hydroxide according to the reaction

$$K_2Ni(CN)_4 + 2HgO + 2H_2O = Ni(OH)_2 + 2KOH + 2Hg(CN)_2.$$

Dilute somewhat with hot water, if necessary, filter, wash with hot water, dry, ignite and weigh as NiO. The ignition should be performed under a hood to avoid the poisonous fumes from the mercury compound present. To determine the impurities in the NiO it may, after weighing, be transferred to a beaker, boiled with water, and then filtered, dried, ignited, and weighed again. Note the loss, probably due to adhering alkali. Now transfer to a porcelain dish, dissolve in nitrohydrochloric acid, evaporate to dryness, moisten with strong hydrochloric acid and then take up in hot water. Filter through a small filter, wash with hot water and then ignite the moist filter and residue and weigh as SiO_2. After deducting the weights of the alkali and silica from

* See also 14, 15, and 18.

the original weight, the balance may be considered as the true weight of the NiO from which the weight of the nickel may be calculated.

10. To determine the cobalt in the filtrate from the nickel, carefully neutralize with nitric acid so as to leave the liquid very slightly alkaline. It must not be acid or strongly alkaline. Now add a solution of mercurous nitrate as long as a precipitate (mercury cobalticyanide) is produced. Filter off the precipitate, wash, dry, and ignite with access of air. Weigh as Co_3O_4. Multiply by 0.7345 to obtain the weight of the cobalt. The degree of oxidation of the cobalt varies somewhat with the temperature of the ignition, and it is, therefore, safer to reduce the oxide by ignition in hydrogen and weigh as metallic cobalt.

11. **Electrolytic Method.** — Proceed as in 7 and continue as described until the zinc has been expelled and the residue dissolved in nitrohydrochloric acid. Evaporate now, on the water-bath, completely to dryness and then dissolve the residue in a little dilute sulphuric acid. Transfer to the beaker in which the electrolysis is to be made, add about 5 grams of ammonium sulphate and 40–60 cc. of strong ammonia and dilute to about 125 cc. with distilled water. The solution is now ready for electrolysis.

The electrolytic apparatus may be the same as described for copper in XIII, 10. See article ELECTROLYSIS, p. 10, for details as to arrangement of apparatus, etc.

12. Insert the electrodes and electrolyse at room temperature with a current density of $ND_{100} = 0.5 - 0.7$ amp. and an electrode tension of 2.8–3.3 volts. The electrolysis is usually finished in from 3 to 4 hours. To test for its completion a little of the solution may be removed with a capillary tube and tested with ammonium sulphide. If no black precipitate is obtained the operation may be considered ended. The solution

should still be strongly ammoniacal. Having broken the circuit, remove the cathode, wash it first with water and then with alcohol (best, absolute), dry at about 100° C. and weigh. The excess over the original weight represents the combined nickel and cobalt.

13. To separate the metals, dissolve them from the electrode with a little strong nitric acid in a small beaker, kept covered as much as possible. Remove and wash the electrode and transfer the solution to a large casserole and evaporate on the water-bath completely to dryness. Now take up in sufficient strong potassium cyanide solution to completely dissolve the residue, but avoiding any great excess, and proceed as described in 9. The cobalt may either be determined as there described or estimated by difference.

14. **Separation of Nickel and Cobalt by Liebig's Potassium Cyanide Method.**—Proceed as described in 7 and 8 until zinc, if present, has been expelled and the nitrohydrochloric acid solution of the residue has been evaporated nearly to dryness to expel the excess of acid. Now dilute sufficiently and neutralize with pure potassium hydroxide, finally adding 5 grams in excess; then add pure potassium cyanide until the precipitate redissolves and the potassium cyanide is in slight excess. To the warm solution add saturated bromine water with constant stirring until the precipitation of the nickel as black nickelic hydroxide is complete. The solution must be kept strongly alkaline throughout the process. Finally, dilute the mixture with cold water and filter, washing the precipitate with hot water.

Dissolve the precipitate in dilute sulphuric acid and determine the nickel electrolytically, as described in 11.

The filtrate contains the cobalt as potassium cobalticyanide. Acidify with dilute sulphuric acid and evaporate as far as possible on the water-bath, then add a little concentrated sulphuric acid and heat the mixture over a free flame until dense white fumes are evolved and effervescence has ceased. The cobalt in the colorless potassium cobalticyanide is thus changed into the rose-red sulphate. Cool, dilute, filter from any silica, and determine the cobalt in the filtrate electrolytically, precisely as described for nickel (11). Or, the filtrate may be treated as follows: Heat to boiling in a porcelain casserole and precipitate the cobalt as black cobaltic hydroxide by the addition of potassium hydroxide and bromine water. Filter through a close filter (such as Schleicher and Schull's No. 589, blue band), dry and ignite. After cooling, extract with water to remove the small amount of alkali always present, then dry the residue and ignite in a current of hydrogen in a Rose crucible, finally weighing the cobalt as metal.

After weighing, dissolve the metal in hydrochloric acid, evaporate to dryness, moisten the residue with hydrochloric acid, dilute and filter off the small residue of silica. Ignite, weigh and deduct this weight from the previous one to obtain the true weight of the cobalt.

15. **Separation of Nickel and Cobalt by the Nitroso-β-Naphthol Method.**—This method is especially suitable for the determination of a small amount of cobalt in the presence of a large amount of nickel, since the cobalt precipitate is very voluminous.

Nitroso-β-naphthol, $C_{10}H_6O(NOH)$, forms with cobalt the compound $Co[C_{10}H_6O(NO)]_3$, cobalti-nitroso-β-naphthol, which is insoluble in hydrochloric acid, while the corresponding nickel compound is soluble.

Proceed as described in 7 and 8 until, after the expulsion of any zinc, the residue has been dissolved in nitrohydrochloric acid. Add a little sulphuric acid and heat over a free flame to

* Ilinske and von Knorre. Ber., 18, 699.

dense white fumes. Cool, dilute, and add 5 cc. of strong hydrochloric acid. Now add a freshly prepared hot solution of nitroso-β-napthol in 50 per cent. acetic acid as long as a precipitate is produced. Allow to settle and then test to see if the precipitation is complete. Digest at the ordinary temperature for a few hours. Filter the precipitate, which consists of the cobalt compound mixed with considerable of the reagent, and wash, first with cold water, then with warm 12 per cent. hydrochloric acid, thoroughly, to remove all the nickel. Finally, wash with hot water until free from acid.

Dry the precipitate and place it in a weighed Rose crucible. Add a little pure crystallized oxalic acid and ignite.* Heat at first very cautiously to avoid loss, and then with the full power of a Bunsen burner. When the carbon of the filter-paper is all consumed, reduce the cobalt to metal by heating in a current of hydrogen. Cool and weigh.

Add sulphuric acid to the filtrate, evaporate to small bulk and then expel the greater part of the acid by heating over a free flame.

The nickel may now be determined as described in 5, or electrolytically (11).

16. Sensitive Test for Nickel.† — α-Dimethylglyoxime is a very sensitive reagent for nickel in solution, and gives a scarlet precipitate, or, with traces of nickel, a yellowish solution from which the red precipitate separates on cooling. The best method of applying this reagent is to remove the excess of acid from the solution to be tested by adding excess of ammonia or sodium acetate, and then to add the powdered dimethyl-

* Von Knorre has subsequently suggested that the oxalic acid may be omitted. Zeit. Angw. Chem., 1893, 264.

† L. Tschugaeff. Ber., 1905, 38, 2520–2522. Jour. Soc. Chem. Ind., XXIV, 941.

glyoxime, and to boil for a short time. Distinct indications are obtained with solutions containing only one part of nickel per 400,000 of water. The reaction is not disturbed by the presence of ten times as much cobalt as nickel, but when the proportion of cobalt is much greater than that of nickel, it is best to add a very large excess of ammonia to the liquid and to shake repeatedly, excess of the dimethyl glyoxime being then added and the solution boiled for a short time. In testing by this method for nickel in such products as commercial cobalt salts, the reaction is manifested by the appearance of a scarlet scum rising up the walls of the test-tube, but it is generally necessary to filter or siphon off the cooling liquid and wash the residue with water; in the presence of nickel this residue is red, but if nickel is absent, it is quite colorless. In this way one part of nickel can be detected when admixed with 5000 parts of cobalt.

17. **Sensitive Test for Cobalt.***—If a concentrated solution of ammonium sulphocyanate is added to a cobaltous solution, the latter becomes a beautiful blue, owing to the formation of ammonium cobaltous sulphocyanate. On adding water the blue color disappears and the pink color of the cobaltous salt takes its place. If, now, amyl alcohol is added (or a mixture of equal parts of amyl alcohol and ether), and the solution is shaken, the upper alcoholic layer is colored blue. This reaction is so sensitive that the blue color is recognizable when the solution contains only 0.02 mg. of cobalt. Nickel salts produce no coloration of the amyl alcohol. If, however, iron is present, the red $Fe(CNS)_3$ is formed, which likewise colors the amyl alcohol, making the blue color (due to cobalt) very indistinct, so that, under some conditions, it can no longer be detected. If, however, some sodium

* Vogel's reaction. From Treadwell's Analytical Chemistry (Hall), Vol. I, p. 137.

carbonate solution is added, the iron will be precipitated as ferric hydroxide, while the blue color produced by the cobalt is unaffected.

18. Separation of Nickel from Cobalt as Nickel Glyoxime, $Ni(C_4H_7N_2O_2)_2$.*—Determine the nickel and cobalt together by electrolysis, as previously described. Dissolve the deposit in nitric acid and replace the latter by evaporation with hydrochloric acid. Dilute the neutral or slightly acid solution so that not more than 0.1 gram of cobalt is present in 100 cc., heat nearly to boiling and add somewhat more than the theoretical amount of an alcoholic solution of dimethyl glyoxime. Now add ammonia cautiously until the solution smells slightly. While the liquid is still hot, filter the precipitate through a Gooch crucible, wash with hot water and dry at 110°–120° C. for 45 minutes. It contains 20.32 per cent. of nickel. The precipitate is red and crystalline. It contains no water of crystallization and sublimes at 250° C. without decomposition.

19. Determination of Nickel in Steel.†—Dissolve about 0.5 gram of the steel in 10 cc. of strong hydrochloric acid, add sufficient nitric acid to completely oxidize the iron, and, if silica separates, add also a little hydrofluoric acid. Finally, add 2 or 3 grams of tartaric acid and dilute the solution to about 300 cc. Carefully test the solution to see if enough tartaric acid is present to prevent any precipitation of iron when the solution is made alkaline with ammonia, and add more tartaric acid if necessary. Now neutralize the solution so as to leave it *slightly* acid, heat nearly to boiling and add 20 cc. of a one per cent. alcoholic solution of dimethyl glyoxime; then very carefully neutralize the slight excess of acid with ammonia, leaving the solution so that it barely smells of this reagent. Filter the hot mixture through a Gooch filter, wash the precipitate with hot water and dry at 110°–120°

---

* Treadwell's Anal. Chem. (Hall), 2d Ed., Vol. II, p. 125.

† *Ibid.*, p. 146. O. Brunck, Stahl und Eisen, 28, 331.

C. for 45 minutes. Weigh as $Ni(C_4H_7N_2O_2)_2$, containing 20.32 per cent. of nickel.

The determination can be made within about two hours. The results are accurate, but may sometimes appear low from the fact that any cobalt present is not determined with the nickel, as it may be with some other methods.

20. Volumetric Method for Nickel.—In the following method* nickel may be accurately determined in the presence of aluminum, ferric iron, magnesium and zinc. Manganese and copper must be removed. Cobalt counts as nickel. If present, it may be detected during the titration by the darkening of the solution. The method is especially applicable to the assay of nickel mattes and German silver. In the case of ores it may be applied to the purified solution after removing iron, manganese, copper, etc., and boiling off any hydrogen sulphide.

The following solutions are required:

Standard silver nitrate, containing about 3 grams of silver per liter. The strength of this solution must be accurately known.

Potassium iodide, 10 per cent. solution.

Potassium cyanide, 22 to 25 grams per liter. This solution must be tested every few days.†

Standardizing the Cyanide Solution. — First, accurately establish the relation of the cyanide to the silver solution. Run into a beaker 3 or 4 cc. of the former, dilute this with about 150 cc. of water, render slightly alkaline with ammonia, and then add a few drops of the potassium iodide. Now carefully run in the silver solution until a faint permanent opalescence is produced, which is finally caused to disappear by the further addition of a mere trace of cyanide. The respective volumes of the silver and cyanide solutions are then read off, and the equivalent

* T. Moore, Chem. News, 72, 92.

† The addition of about 5 grams of potassium hydroxide per liter is said to increase its permanency

in cyanide of 1 cc. of silver solution calculated. Now calculate the metallic silver value of 1 cc. of the cyanide solution and multiply this by 0.27193 to obtain the nickel value.

Titration of a Nickel Solution. — The nickel solution (containing not more than about 0.1 gram of nickel) must have sufficient free acid present to prevent the formation of any precipitate on the subsequent addition of ammonia to alkaline reaction; if this is not the case, a little ammonium chloride may be added. Make distinctly alkaline with ammonia, add a few drops of the potassium iodide solution and dilute to 150 or 200 cc. A few drops of the silver solution are now run in and the solution stirred to produce a uniform turbidity. Cool the mixture to at least 20° C. and it is ready to be titrated with the potassium cyanide solution, which is added slowly and with constant stirring until the precipitate wholly disappears; a few extra drops are added, after which the beaker is placed under the silver nitrate burette, and this solution gently dropped in until a faint permanent turbidity is again visible; this is now finally caused to dissolve by the mere fraction of a drop of the cyanide. A correction must now be applied for the excess of cyanide added, by noting the amount of silver solution employed and deducting its value in cyanide solution from the amount of the latter used. The amount of cyanide solution actually consumed by the nickel is thus arrived at, from which the percentage of nickel is calculated.

In the presence of aluminum, magnesium or ferric iron, they may be kept in solution with either citric acid, tartaric acid or sodium pyrophosphate. In the presence of zinc, use sodium pyrophosphate. The action of iron is somewhat deceptive, as the solution, once cleared up, often becomes troubled again on standing for a minute; should this occur, a further addition of cyanide must be given until the liquid is rendered perfectly limpid. The temperature of the solution should never exceed 20° C.: above this the results become irregular. The amount of free ammonia also has a disturbing influence; a large excess hinders or entirely prevents the reaction; the liquid should, therefore, be only slightly but very distinctly alkaline. The potassium cyanide should be as pure as possible. The most hurtful impurity is sulphur, which causes a darkening of the solution owing to the formaion of the less readily soluble silver sulphide. To get rid of the sulphur impurity it is necessary to thoroughly agitate the cyanide liquor with lead oxide, or, what is far preferable, bismuth oxide.

In the hands of Mr. Moore, the above method, after many thousands of determinations, has been found to be more accurate and reliable than either the electrolytic or gravimetric methods.

21. Modification of the Above Method, Applied to the Determination of Nickel in Nickel Steel.*—The determination can be made with speed and accuracy, even in the presence of iron, manganese, chromium, zinc, vanadium, molybdenum and tungsten.

Requirements. The potassium cyanide solution should be about equivalent to a tenth-normal silver solution, and is prepared by dissolving 13.5 grams of pure potassium cyanide and 5 grams of potassium hydroxide in water and diluting to 1 liter.

The silver nitrate solution is made exactly tenth-normal and is prepared by dissolving 8.495 grams of $AgNO_3$ in water and diluting to exactly 500 cc. 1 cc. of this solution is equivalent to 0.01302 grams KCN, or to 0.002934 grams·Ni. It is used for standardizing the potassium cyanide solution, and in the analysis itself.

The potassium iodide solution contains 2 grams KI in 100 cc.

Determination. Dissolve 1 gram of the steel in a casserole

* Treadwell's Anal. Chem. (Hall), 2d Ed., p. 656. Numerous references are given.

with 10 to 15 cc. of nitric acid (sp. gr. 1.20), adding a little hydrochloric acid if necessary. After the steel has dissolved, add 6 to 8 cc. of sulphuric acid (1:1) and evaporate until fumes of the latter begin to come off. Cool the residue, add 30 to 40 cc. of water and heat until the anhydrous ferric sulphate has all dissolved. Transfer the solution to a 400-cc. beaker, filtering if necessary, and add 13 grams of sodium pyrophosphate dissolved in 60 cc. of water at about 60° C. The pyrophosphate solution must not be boiled, as this causes the formation of normal phosphate. The addition of the sodium pyrophosphate produces a heavy white precipitate of ferric pyrophosphate. Cool the liquid to room temperature and add dilute ammonia (1:1), drop by drop, while stirring constantly, until the greater part of the precipitate has dissolved and the solution has assumed a greenish tinge. At this point it should react alkaline toward litmus, but should not smell of free ammonia. Now heat the solution gently, while stirring, and the remainder of the pyrophosphate will dissolve, giving a perfectly clear light green solution. If the ammonia is added too fast, or the solution is not carefully stirred, a brownish color is likely to result, but this can usually be overcome by carefully adding a few drops of dilute sulphuric acid. Cool the clear solution to room temperature and add 5 cc. of the standard silver nitrate solution together with 5 cc. of the potassium iodide. Now titrate the solution with the potassium cyanide, adding it until the precipitate of silver iodide has disappeared. Finish the titration by adding just enough more of the silver nitrate to again produce a slight turbidity. Before calculating the percentage of nickel, make a correction for the amount of silver nitrate added.

Remarks.—The results obtained by the potassium cyanide titration of nickel are said to be very satisfactory. It can be carried out in the presence of most of the other elements of the ammonium sulphide group.* If copper is present in amounts not exceeding 0.4 per cent., the copper will replace almost exactly three-quarters of its weight of nickel. In case chromium is present, the dark color due to the presence of chromic salts may be obviated by adding to the original sulphuric acid solution a two per cent. solution of potassium permanganate until a slight permanent precipitate of manganese dioxide is obtained, whereby the chromium is oxidized to chromic acid. The solution is filtered. concentrated in a 400 cc. beaker to about 60 cc., then treated with sodium pyrophosphate, as described above. The method is not applicable in the presence of cobalt, but when the amount of the latter does not exceed one-tenth the amount of nickel present, the titration can be carried out successfully and the results represent the amount of nickel and cobalt present.

22. **Electrolytic Method in General Use at Cobalt, Ont., Canada.**†—Treat 1 gram of the ore in a 250-cc. beaker with 10 cc. of strong nitric acid and 5 cc. of strong hydrochloric acid. When decomposition is complete add 15 cc. of strong sulphuric acid and boil until the sulphuric acid is fuming strongly. Allow to cool, take up with water, add 5 cc. of strong hydrochloric acid to assist solution and boil gently for a few minutes. Remove from the heat and pass hydrogen sulphide through the hot solution for ten to fifteen minutes, then heat again and once more pass in hydrogen sulphide until all the arsenic is precipitated and the supernatant liquid is clear. Filter, washing the precipitate six or seven times with hot water. Boil the filtrate to expel the hydrogen sulphide and reduce to a volume of about 150 cc. Filter and wash as before. Heat the filtrate to boiling and add 15 to 20

---

* Instead of using sodium pyrophosphate to prevent the interference of iron and other metals, many chemists use citric or tartaric acid.

† Communicated by Mr. B. C. Lamble, chemist at the La Rose Mines, Ltd., Cobalt.

cc. of hydrogen peroxide to effect oxidation. After boiling a short time, remove from the heat and allow to cool somewhat and then separate the iron as basic acetate as follows: Make the solution just alkaline with ammonia and then reacidify slightly with hydrochloric acid. Add 15 grams of solid sodium acetate (crystals) and boil. Filter off the iron precipitate, washing with hot water, and place the filtrate over the heat to concentrate. Dissolve the precipitate in hot dilute hydrochloric acid, dilute to about 150 cc. and repeat the precipitation as just described. Unite the two filtrates, boil down to about 100 cc. and filter. Make the filtrate alkaline with ammonia and add about 20 cc. in excess. Electrolyze the hot solution (which should have a volume of 125 to 130 cc.), keeping it hot during the electrolysis and highly ammoniacal. (See **12.**) When the operation is apparently finished (in from 2½ to 3½ hours), test for nickel and cobalt on a spot-plate with ammonium sulphide. Wash the cathode first with water, then with alcohol, dry at about 100° C., and weigh. This gives the combined nickel and cobalt.

Dissolve the deposit in a 250-cc. beaker in strong nitric acid and wash off the cathode with hot water. Boil or evaporate the solution to a syrup. Add about 20 cc. of water and then add a solution of potassium hydroxide until a precipitate of cobalt hydroxide (blue) remains undissolved. Now acidify with acetic acid and add a few cubic centimeters in excess; then add about 25 grams of potassium nitrite and stir until dissolved. Boil gently for about twenty minutes, then dilute with water until the beaker is about five-sixths full and allow to stand over night. Decant and filter the cobalt precipitate and wash it a few times with ice-cold water, or water containing about five per cent. of potassium nitrite and barely acidified with acetic acid. Dissolve the precipitate in hot dilute sulphuric acid and boil down the solution until the sulphuric acid is fuming. Cool, dilute with water, heat to boiling and filter. Make the filtrate alkaline with ammonia, having about 20 cc. in excess, dilute to about 125 cc. and electrolyze as described above. This gives the cobalt, and the nickel is determined by the difference between this weight and that of the combined nickel and cobalt.

NOTE.—I would recommend an 8-oz. flask instead of a beaker for the decomposition, and 7 cc. of sulphuric acid instead of 15 cc. Any globules of separated sulphur may be completely volatilized by allowing the sulphuric acid to boil very gently for some time after coming to fumes. In electrolyzing, Mr. Lamble dilutes the solution to about 300 cc. and employs a gauze cathode.

CHAPTER XXII.

PHOSPHORUS.

(For phosphorus in tungsten ores, see Appendix.)

1. **Gravimetric Method for Iron Ores.***—Take 1.63 grams of the finely ground ore. Dissolve by boiling gently for 20 minutes with 40 cc. of strong hydrochloric acid (1.20 sp. gr.) in a covered 8-oz. beaker. Dilute with 20 cc. of water and filter, washing with water. Receive the filtrate in an 8-oz. beaker. Evaporate the filtrate to dryness on a sand-bath or hot plate. While the filtrate is evaporating ignite the filter and insoluble residue in a platinum crucible. After the paper is burned off, break up any lumps with a platinum rod and then ignite again at a red heat for about 2 minutes. Cool, and transfer the ignited residue to the beaker in which the filtrate is evaporating. Any phosphorus that remained with the insoluble residue is thus rendered soluble by the ignition and recovered.† After evaporation, add to the residue in the beaker 25 cc. of strong nitric acid (sp. gr. 1.42), cover the beaker and boil to about 12 cc. Now dilute with 12 cc. of water and filter, washing with water. Receive the filtrate in an 8-oz. flask. The total bulk should not exceed 50 cc. Heat to 40°–45° C., add 60 cc. of molybdate solution,‡ previously

* Mainly the method of John S. Unger, Trans. Engineers Soc. of Western Penn., 1896.

† Mixer and Dubois. Jour. Am. Chem. Soc., XIX, p. 614.

‡ See below, 7.

filtered and heated to 40°–45° C., stopper the flask and shake for 5 minutes. Allow to stand in a warm place for 15 minutes and then filter through a Gooch crucible that has been previously dried at 110° C. and weighed. Wash with a 2 per cent. nitric acid solution till free from iron, and then twice with 95 per cent. alcohol. Dry 20 minutes at 110° C. and weigh. The dried residue contains 1.63 per cent. of phosphorus, therefore each milligram found corresponds to 0.001 per cent. of phosphorus in the ore.

2. **Error Caused when Titanium is Present.**—Blair * notes the fact that if a solution of ferric chloride containing titanic and phosphoric acids is evaporated to dryness, a compound of TiO_2, P_2O_5, and Fe_2O_3 is formed which is completely insoluble in hydrochloric acid (and presumably in nitric acid also). When titanium is present in the material analyzed and such a residue is obtained, on filtering the nitric acid solution after evaporation to dryness, it may be treated as follows:

3. Dry the residue and burn off the filter paper by ignition in a platinum crucible. Moisten the cool residue with a few drops of strong sulphuric acid and add sufficient hydrofluoric acid to dissolve the silica. Evaporate cautiously and heat until all the sulphuric acid is driven off. Cool, add 2 or 3 grams of sodium carbonate and fuse the mixture. Dissolve the melt in hot water, filter and wash. Receive the filtrate in a 6-oz. flask and acidify with nitric acid. The total bulk should not exceed 50 cc. Heat to 40°–45° C., add 25 cc. of molybdate solution, previously filtered and heated to the same temperature, and finish as described above for the main solution. The same filter, if desired, may be used for both filtrations of the yellow precipitate.

4. The presence of titanium in an ore, if much is present,

---

* Chem. Anal. of Iron, 3d Ed., p. 86.

may be recognized by the peculiar milky appearance of the solution when it is diluted before filtering off the insoluble matter, and by the tendency of the latter to run through the filter when washed with water. In cases of uncertainty it is best to assume that titanium is present and proceed with the residue as described above. If the residue left after filtering the aqueous solution of the melt of 3 be dissolved in dilute hydrochloric acid, the solution placed in a test-tube and metallic zinc added, the liquid will become first colorless, from the reduction of the ferric iron, then, if titanium be present, pink or purple and finally blue, from the formation of Ti_2O_3.

5. **Method for Steel.**—Dissolve 1.63 grams of the steel, in a 6-oz. Erlenmeyer flask, in 30 cc. of nitric acid of about 1.20 sp. gr. Evaporate by boiling over a naked flame to 15 cc., add to the boiling solution 20 cc. of chromic acid solution (see **8** below) and again evaporate to 18 cc. Remove from the heat and wash down the sides of the flask with from 5–7 cc. of water and cool to 40°–45° C. Proceed from this point as directed above for ores (1). As in the case of ores, each milligram of yellow precipitate found corresponds to 0.001 per cent. of phosphorus in the steel.

6. **Method for Pig Iron.**—Dissolve 1.63 grams, in a 4½-inch covered evaporating-dish, in 40 cc. of strong nitric acid (sp. gr. 1.42). When action has ceased, remove the cover and evaporate to dryness on a sand-bath. Place the dish and dry residue over a burner and heat cautiously with the naked flame until the mass ceases to evolve red fumes. Allow to cool, add 25 cc. of strong hydrochloric acid (sp. gr. 1.20), cover with a watch-glass and boil to 10 cc. Add cautiously 25 cc. of strong nitric acid (sp. gr. 1.42) and boil to 12 cc. Remove from the heat, rinse off the watch-glass and wash down the sides of the dish, using about 12 cc. of water, and then filter through an 11-cm.

filter into an 8-oz. flask, washing with water. The solution should not exceed 50 cc. in bulk. Heat to 40°–45° C. and finish as described above for ores (1). Each milligram of yellow precipitate found corresponds to 0.001 per cent. of phosphorus in the pig iron.

7. **Molybdic Acid Solution.***—Mix 100 grams of molybdic acid acid to a paste with 265 cc. of water. Add 155 cc. of strong ammonia water (sp. gr. 0.90) and stir until all is dissolved. To this solution add 66 cc. of strong nitric acid (sp. gr. 1.42), stir well and then set aside for an hour. In another vessel make a mixture of 395 cc. of strong nitric acid and 1100 cc. of water. Finally, pour the first solution into the second, stirring constantly. Allow to stand for 24 hours before using.

8. **Chromic Acid Solution.** — Dissolve 30 grams of pure chromic acid in 2 liters of strong nitric acid (sp. gr. 1.42), warming gently. This solution must be made up fresh at least every two weeks.

9. **Volumetric Method for Iron Ores, Steel, etc.**—Start with 2 grams of material and, according to its nature, proceed by one of the above-described methods until the yellow precipitate is finally obtained on the filter, which need not be weighed. Wash with a solution of acid ammonium sulphate, prepared by adding 15 cc. of strong ammonia (sp. gr. 0.90) and 25 cc. of strong sulphuric acid (sp. gr. 1.84) to 1 liter of water. Wash until 2 or 3 cc. of the wash-water give no brown reaction for molybdenum when tested with a drop of ammonium sulphide. Place the flask in which the precipitation was made under the funnel and dissolve the ammonium phosphomolybdate on the filter by pouring over it a warm mixture of 5 cc. of strong ammonia water and 20 cc. of water. Wash the filter with water until

---

* See 12.

the filtrate measures 75 cc. Now add 5 grams of pulverized pure zinc, 100-mesh, pouring it through a funnel to prevent any zinc from clinging to the sides of the flask. Then add, cautiously, 15 cc. of strong sulphuric acid, best by letting it run from a glass-stoppered burette. It is a good plan, although not absolutely necessary, to close the flask with a rubber stopper carrying a glass tube bent twice at right angles, the outer arm dipping into a beaker containing a saturated solution of sodium acid carbonate. Allow the flask and contents to stand for about 30 minutes, when, if all action has ceased and there is no appreciable residue from the zinc, the solution is ready to titrate with permanganate. In case of a residue (lead, etc.), that might have a reducing action on the permanganate, filter the solution through a rapid-running filter* and wash with luke-warm water receiving the filtrate in a flask. The solution should be of a green color. If brown, the result will come low. Bring the liquid to a temperature of about 40° C. and titrate with standard potassium permanganate solution. The titrated solution gradually becomes colorless and then, at the end, the usual pink tinge is obtained.

10. A blank test should be made by adding to another flask 70 cc. of water, 5 cc. of strong ammonia, 5 grams of the 100-mesh zinc and finally 15 cc. of strong sulphuric acid, and treating this flask precisely like the other. The amount of permanganate used by this test should be deducted from the amount used in the analysis of the substance.

The iron value of the permanganate solution multiplied by 0.01541 will give the phosphorus value.

This may be determined as follows:

11. When molybdic acid is reduced to Mo_2O_3 the latter requires the same amount of oxygen for reoxidation as is required to oxidize 6FeO to the ferric state. Therefore, 3Fe (ferrous) are equivalent to MoO_3, or, 167.7 parts by weight of Fe are

* A filter of absorbent cotton and the procedure described in VIII, 12, is best.

equivalent to 144 parts of MoO_3. Accordingly, the iron factor of the permanganate multiplied by 0.8586 will give the MoO_3 factor. Now the formula of the ammonium phosphomolybdate is $(NH_4)_3 12MoO_3.PO_4$, therefore 1728 parts by weight of MoO_3 correspond to 31 parts of P, or, 1 part of MoO_3=0.01795 parts of P. Thus by multiplying the iron factor of the permanganate by 0.8586, and the product by 0.01795, the phosphorus factor is obtained, or, the Fe factor multiplied by 0.01541 gives the P factor.

12. Note Regarding Molybdic Acid.—The substance sold as molybdic acid frequently consists to a greater or less extent of alkali molybdate. Such material may be used for phosphorus determinations with fairly good results if the percentage of molybdic acid is determined and the amount of substance taken in making up the molybdate solution regulated accordingly, so that the full quantity of molybdic acid shall be present. The determination may be made by weighing 0.1 gram, dissolving in a mixture of 70 cc. of water and 5 cc. of strong ammonia, adding 5 grams of 100-mesh zinc and 15 cc. of strong sulphuric acid, and finally titrating the reduced solution precisely as in a phosphorous determination (9).

The iron factor of the permanganate multiplied by 0.8586 will give the MoO_3 factor (10).

A yellow molybdate solution indicates the presence of silica, and a phosphorus determination made with such a solution is likely to come high owing to ammonium silicomolybdate being dragged down with the phosphomolybdate. If about 30 milligrams of microcosmic salt dissolved in a little water be added per liter of yellow molybdate solution, the mixture agitated and then allowed to settle 24 hours, the solution will become colorless and fit for use.

13. Method for Limestone.—Treat 20 grams in a large covered beaker with sufficient strong hydrochloric acid to effect solution. Add the acid cautiously, a little at a time, with agitation, to avoid frothing over the beaker. When enough acid has been added and effervescence has ceased, warm gently if necessary to dissolve any separated iron compounds. Solution being finally as complete as possible, dilute sufficiently with cold water and filter, washing filter and residue with cold water. To the filtrate add a few drops of ferric chloride solution (not necessary, of course, if the solution already contains sufficient iron) and then ammonia until the solution is alkaline to litmus paper. If the precipitate, containing the phosphorus as ferric phosphate, is not decidedly reddish-brown in color,* acidulate again with hydrochloric acid, add more ferric chloride solution and then ammonia once more to alkaline reaction. Now add acetic acid to decided acid reaction. Boil for a few minutes, filter, wash once with hot water and then dissolve in warm dilute hydrochloric acid on the filter, receiving the solution in the beaker in which the precipitation was made. Reserve this solution for the present.

14. The original insoluble residue may contain phosphorus. Ignite it in platinum until the carbon of the filter is all burned off, break up any lumps with a platinum rod and then ignite at a red heat for 2 or 3 minutes. Cool, moisten with water, add a little hydrochloric acid and warm to dissolve the soluble matter. Dilute and filter through the last filter used above, receiving the filtrate in the beaker containing the reserved solution. Dilute this solution and reprecipitate exactly as before with ammonia and acetic acid. Dissolve this precipitate on the filter in warm dilute hydrochloric acid, receiving the solution in a small beaker and washing the filter with hot water. Evaporate the solution in the beaker almost to dryness to expel the excess of hydrochloric

* Ferric phosphate is white and the red color indicates the necessary excess of iron precipitated as ferric hydroxide.

acid, and then add to it a filtered solution of 5 or 10 grams of citric acid (according to the size of the ferric precipitate) dissolved in 10–20 cc. of water. Next, add 5 to 10 cc. of magnesia mixture * and enough ammonia to make the liquid faintly alkaline. Stand the beaker in cold water until perfectly cold and then add to the solution one-half its volume of strong ammonia and stir well. When the precipitate of NH_4MgPO_4 has begun to form, stop stirring and allow the beaker to stand in cold water for 10 or 15 minutes, then stir vigorously several times at intervals of a few minutes, and finally allow the mixture to stand over night. Filter on a small ashless filter and wash with a mixture of 1 part strong ammonia and 2 parts water and containing also 2.5 grams of ammonium nitrate in 100 cc.

15. Dry the filter and precipitate and ignite them in a weighed platinum crucible, first at a very low temperature so as to carbonize the filter without decomposing the precipitate. Now break up the residue with a platinum rod and then heat at a gradually increasing temperature to the full power of a Bunsen burner and continue the ignition until the residue is perfectly white. Cool and weigh. Now fill the crucible half full of hot water, add from 5 to 20 drops of strong hydrochloric acid and heat until the precipitate has dissolved. Filter off on another small ashless filter any silica or ferric oxide that may remain, ignite and weigh. The difference between the two weights is the weight of the $Mg_2P_2O_7$, which, multiplied by 0.27837, will give the weight of the phosphorus.

* *Magnesia mixture.*—Dissolve 110 grams of crystallized magnesium chloride ($MgCl_2 + 6H_2O$) in water and filter. Dissolve 280 grams of ammonium chloride in water, add a little bromine water, and a slight excess of ammonia, and filter. Add this solution to the solution of magnesium chloride, add enough ammonia to impart a decided odor, dilute to about 2 liters, shake vigorously from time to time, allow to stand for several days, and filter into a small bottle as required for use. Ten cc. of this solution will precipitate about 0.15 gram of P_2O_5. Blair, Chem. Anal. of Iron.

CHAPTER XXIII.

POTASSIUM AND SODIUM.

ALMOST the only occasions when the technical metallurgical chemist has to make quantitative determinations of potassium and sodium are in the analyses of silicates, such as clays, etc. Descriptions of methods will, therefore, be confined to such as apply to these substances.

1. **Method of J. Lawrence Smith.*** —*Principle.*—The substance is heated with a mixture of 1 part ammonium chloride and 8 parts calcium carbonate. By this means the alkalies are obtained in the form of chlorides, while the remaining metals are for the most part left behind as oxides, and the silica is changed to calcium silicate, as represented by the following equation:

$$2KAlSi_3O_8 + 6CaCO_3 + 2NH_4Cl = 6CaSiO_3 + 6CO_2 + Al_2O_3 + 2KCl + 2NH_3 + H_2O.$$

The alkali chlorides together with the calcium chloride can be removed from the sintered mass by leaching with water, while the other constituents remain undissolved.

Preparation.—The ammonium chloride necessary for the determination is prepared by subliming the commercial salt; the calcium carbonate by dissolving the purest calcite obtainable

* This description of Smith's method is taken from Treadwell's Analytical Chemistry, Hall, Vol. II.

in hydrochloric acid and precipitating with ammonia and ammonium carbonate. This last operation is performed in a large porcelain dish. After the precipitate has settled, the clear solution is poured off and the precipitate is washed by decantation until free from chlorides. The product thus obtained contains traces of alkalies, but the amount present is determined once for all by a blank test and a corresponding deduction made from the results of the analysis; it is usually sodium chloride and amounts to 0.0012–0.0016 gram for 8 grams calcium carbonate. The decomposition was performed by Smith in a finger-shaped crucible about 8 cm. long and with a diameter of about 2 cm. at the top and 1½ cm. at the bottom. Such a crucible is suitable for the decomposition of about 0.5 gram of the mineral. A larger quantity can be analyzed in an ordinary platinum crucible.

Filling the Crucible.—About 0.5 gram of the mineral is mixed with an equal quantity of sublimed ammonium chloride by trituration in an agate mortar, then 3 grams of calcium carbonate are added and intimately mixed with the former. The mixture is transferred to a platinum crucible with the help of a piece of glazed paper, and the mortar is rinsed with 1 gram of calcium carbonate, which is added to the contents of the crucible.

The Ignition.—The covered crucible is placed in a slightly inclined position and gradually heated over a small flame until no more ammonia is evolved * (this should take about 15 minutes), then the temperature is raised until finally the lower three-fourths (and no more) of the crucible are brought to a dull red heat, and this temperature is maintained for 50 or 60 minutes. The crucible is then allowed to cool and the sintered cake can

* During this part of the operation the heat should be kept so low that ammonium chloride does not escape. The latter is dissociated into ammonia and hydrochloric acid by the heat, and the acid unites with the calcium carbonate to form calcium chloride.—(Hall.)

usually be removed by gently tapping the inverted crucible. Should this not be the case, it is digested a few minutes with water, which serves to soften the cake so that it can be readily washed into a large porcelain, or, better, platinum dish. The covered dish is heated with 50–75 cc. of water for half an hour, replacing the water lost by evaporation, and the large particles are reduced to a fine powder by rubbing with a pestle in the dish. The clear solution is decanted through a filter and the residue is washed four times by decantation, then transferred to the filter and washed with hot water until a few cubic centimeters of the washings give only a slight turbidity with silver nitrate. To make sure that the decomposition of the mineral has been complete, the residue is treated with hydrochloric acid; it should dissolve completely, leaving no trace of undecomposed mineral.

Precipitation of the Calcium.—The aqueous solution is treated with ammonia and ammonium carbonate, heated and filtered. As this precipitate contains small amounts of alkali, it is redissolved in hydrochloric acid and the precipitation with ammonia and ammonium carbonate is repeated. The combined filtrates are evaporated to dryness in a porcelain or platinum dish, and the ammonium salts are removed by careful ignition over a moving flame.* After cooling, the residue is dissolved in a little water and the last traces of calcium are removed by the addition of ammonia and ammonium oxalate. After standing 12 hours, the calcium oxalate is filtered off and the filtrate is received in a weighed platinum dish, evaporated to dryness, and gently ignited. After cooling, the mass is moistened with hydrochloric acid in order to transform any carbonate into chloride, the evaporation and ignition are repeated, and the weight of the contents of the dish is determined; this represents the amount

* Before igniting, it is well to heat the contents of the dish in a drying-oven at 110°. By this means there is no danger of loss by decrepitation.—(Hall.)

of alkali chloride present. To determine potassium, the residue is dissolved in water, and the potassium is precipitated as potassium platinic chloride, as described in §§ 3, 4. The sodium is determined by difference.

2. **Hydrofluoric Acid Method.**—Treat 1 gram of the finely powdered silicate in a platinum dish with pure strong hydrochloric and hydrofluoric acids until decomposition is complete. It is best to add the hydrochloric acid first, to prevent spattering, and then about an equal amount of hydrofluoric acid and warm gently. Finally, evaporate to dryness on the water-bath, then add a little dilute hydrochloric acid and evaporate again to dryness. Dissolve the residue in hot water, add 5 cc. of a saturated solution of barium hydroxide, and heat to boiling Allow to settle a short time and then test the clear liquid with a little more barium hydroxide solution to be certain that enough has been added. When the precipitation is complete, filter and wash well with hot water. Heat the filtrate to boiling and add a little ammonia and ammonium carbonate to complete the precipitation of calcium, barium, etc., and, after allowing to stand a short time on the water-bath, filter and wash the precipitate thoroughly with hot water. Evaporate the filtrate to dryness in platinum or porcelain and then expel the ammonium salts by a heat just below redness. Take up the cool residue in a little hot water, add a few drops of ammonia, a drop or two of strong ammonium carbonate solution and a few drops of ammonium oxalate solution. After allowing to stand a short time on the water-bath, set aside for a few hours and then filter, washing with hot water. Evaporate the filtrate to dryness on the water-bath and heat the residue to dull redness until all ammonium salts are expelled and it has become nearly or quite white. Cool, dissolve in a very small amount of water and filter into a weighed platinum dish. Add a few drops of hydrochloric acid and evaporate to

dryness on the water-bath. Heat the residue to dull redness, cool in a desiccator and weigh as mixed potassium and sodium chlorides. Repeat the heating until a constant weight is obtained. Dissolve the chlorides in a small amount of water. If a residue remains undissolved the separation must be repeated until the chlorides finally obtained are entirely soluble.

Proceed with the solution of the chlorides as described in 3. or 6.

SEPARATION OF POTASSIUM AND SODIUM.

3. **Direct Method.**—It is assumed that the two metals are in aqueous solution as chlorides in a platinum or porcelain evaporating dish, the weight of the mixed chlorides being known. Add a slight excess of hydrochlorplatinic acid (H_2PtCl_6) and evaporate to dryness at a low temperature on a water-bath in which the water is not boiling. The drying is to dehydrate the sodium platinic chloride and render it more soluble in alcohol. It is a good plan to have the platinum solution contain 10 per cent. of platinum, i.e., 1 gram in every 10 cc., and, in order to use the right quantity and avoid unnecessary excess, make the following calculation:

Assume that the mixed chlorides consist entirely of NaCl. Call their weight *a*. Then from the formula $Na_2PtCl_6+6H_2O$ we find that 46.10 parts of Na require 194.6 parts of Pt. 46.1 parts of Na correspond to 117 parts of NaCl; therefore we have the proportion

$$117:194.6=a:\text{Pt required.}$$

Solving this and multiplying the result by 10 we arrive at the number of cubic centimeters of platinum solution required. Always add 0.3–0.4 cc. in excess.

It is necessary to convert all the sodium as well as the potassium to the platinum compound, as otherwise the undecomposed sodium chloride, being insoluble in absolute alcohol, would contaminate the potassium platinic chloride; the calculation is therefore made as above.

4. After the evaporation add to the cool residue a few cubic-centimeters of absolute alcohol (methyl alcohol is the best) and thoroughly disintegrate the solid mass with a stirring-rod or platinum spatula. Decant the clear liquid through a filter moistened with alcohol, and then repeat the stirring with fresh portions of alcohol until a perfectly colorless filtrate is obtained and the remaining salt in the dish is pure golden yellow with no intermixed orange-colored particles of Na_2PtCl_6. The latter compound is soluble in the alcohol while the corresponding potassium salt is not. Transfer the washed residue to the filter, allow to drain completely and then dry in an air-bath at 80°–90° C. When dry, carefully transfer as much of the precipitate as possible to a watch-glass. Replace the filter in the funnel and dissolve the adhering precipitate (and also any in the original dish) by washing with a little hot water, receiving the filtrate in a weighed platinum dish or crucible. Evaporate the solution to dryness at a low temperature on the water-bath and then add the precipitate on the watch-glass. Dry the whole at 160° C. and weigh as K_2PtCl_6. Multiply this weight by 0.30561 to obtain that of the KCl. Although this factor is correct according to the less recent values of the atomic weights, the factor 0.30712 is the correct one according to the latest values. This figure, however, does not actually give results so near to the truth as the older factor. This is due to the fact that our assumption as to the formula K_2PtCl_6 is not quite correct. Changes are produced by the evaporation which are compensated for by the use of the old factor.

Having determined the weight of the KCl in the mixed chlorides, the difference from the total weight is the NaCl.

Multiply the weight of the KCl by 0.632 to obtain the weight of the K_2O.

Multiply the weight of the NaCl by 0.5308 to obtain the weight of the Na_2O.

5. Indirect Method.—This method is accurate only when both bases are present in considerable amount.

The metals must first be obtained as chlorides by one of the preceding methods and the combined weight noted. Now determine the percentage of chlorine in the mixture (see XI, 1) and then apply the following rule:

Subtract 47.52 from the percentage of chlorine in the mixture. Divide the remainder by 0.1307. The result is the per cent. of sodium chloride in the mixture.

This rule is determined as follows: Pure NaCl contains 60.59 per cent. of Cl. Pure KCl contains 47.52 per cent. of Cl. The difference between the two percentages is 13.07. This 13.07 per cent. excess represents, then, 100 per cent. of NaCl, therefore every 0.1307 per cent. above 47.52 represents 1 per cent. of NaCl; hence the above rule.

6. Separation of Potassium and Sodium by the Perchlorate Method.—It is assumed that the two metals exist as chlorides in a platinum or porcelain dish, the weight of the mixed chlorides being known. Add just enough water to dissolve the salts, then 5 cc., or more, of 20 per cent. perchloric acid and evaporate to dryness on a hot-plate and until all fumes of perchloric acid have disappeared. If the heavy fumes fail to appear, cool, again take up in a little water, add 5 cc. more of perchloric acid and repeat the evaporation. Two evaporations, each showing the heavy fumes, are necessary, the fumes being entirely expelled each time. Finally, allow to become completely cool and then continue as described under Potash, 2, in the Appendix. Multiply the K_2O found by 2.887 to obtain the KCl. Subtract this from the weight of the mixed chlorides to obtain the weight of the NaCl. Multiply this by 0.5308 to obtain the Na_2O.

CHAPTER XXIV.

SILICA.

1. In the valuation of ores in the West it is customary to designate as "silica" the insoluble residue remaining after certain conventional treatments with acids. Precautions are taken to remove lead compounds and sometimes other efforts at purification are made, so that the final insoluble residue, or so-called "silica," shall consist either of fairly pure silica or a mixture of silica and undecomposed silicates.

2. **Insoluble Residue or "Silica" in Ores, etc.***—Weigh 0.5 gram of the ore into a 4-oz. Erlenmeyer flask. The choice of acids for the decomposition will depend upon the apparent nature of the ore. Oxidized ores, especially when they contain much iron or manganese, should first be treated with hydrochloric acid alone, say 10 or 15 cc., and the mixture warmed very gently until decomposition and solution are as complete as possible. Actual boiling should be avoided, as the acid is thereby weakened and rendered less effective. If sulphides finally remain undissolved add a few cubic centimeters of strong nitric acid and heat the mixture again.

If the ore is largely a sulphide it may frequently be attacked at once with 10 cc. of strong nitric acid and the mixture warmed gently until the first violent action has somewhat subsided. Then add 5 cc. of strong hydrochloric acid and continue the warming to complete the decomposition if necessary.

---

* See 21.

In the case of very heavy sulphides, where the separated sulphur after treatment with nitric acid appears to be still impure, it may be advisable to add small portions of potassium chlorate from time to time, together with more nitric acid if necessary, and continue the heating until the sulphur is either entirely oxidized or sufficiently purified. Then add 5 cc. of strong hydrochloric acid, very cautiously, to avoid too violent action with the undecomposed chlorate, and again heat gently.

3. Having obtained a sufficiently complete decomposition by any of the above methods, or modifications of them suggested by the nature of the ore in hand, the solution is to be boiled or evaporated to complete dryness. This may be done either slowly, by gentle heating on a hot plate or water-bath, or rapidly, by manipulating the flask over a small free flame.

When the mixture has become dry, continue the heating for 5 or 10 minutes, or more, at a temperature of about 150° C., in order to dehydrate any gelatinous silicic acid present. Finally, after cooling somewhat, add 10 cc. of strong hydrochloric acid and warm until solution is as complete as possible. Now add 40 cc. of water and about 5 grams of ammonium chloride. The latter is to hold lead salts in solution and may of course be omitted when no lead is present. Heat to boiling and see that everything soluble is properly dissolved. Filter through a 9-cm. filter and wash well with hot water. It is best either to have the water slightly acidulated with hydrochloric acid, or to wash at least once or twice with acidulated water, to remove from the filter any insoluble iron compound due to the hydrolysis of ferric chloride. Any residue adhering in the flask should be dislodged with a rubber-tipped glass rod and rinsed into the filter. Ignite and weigh.

4. It is usually sufficient to place the filter and contents, without drying, in a small clay "annealing-cup" and ignite in a muffle or over a Bunsen burner. When cold shake and brush the

ignited residue upon the scale-pan and weigh. The weight of the filter-ash may ordinarily be neglected. As 0.5 gram of ore was taken, the weight in centigrams multiplied by 2 will give the percentage of "silica."

5. When the methods above described fail to effect a fairly complete decomposition of the ore, sulphuric acid may be tried in addition. Proceed in the usual way with hydrochloric and nitric acids until decomposition is about as complete as possible, and then add 5 cc. of strong sulphuric acid and heat (best over a free flame) until the white fumes are coming off copiously. It is best to continue the heating until most of the sulphuric acid is expelled. Cool, dilute with about 40 cc. of water, add 5 cc. of strong hydrochloric acid and 5 grams of ammonium chloride, and boil to effect solution of everything soluble. Finally, filter, wash, ignite, and weigh as described above.

If an ore contains silicates that are gradually but appreciably decomposed by acids, it is evident that the amount of insoluble residue obtained will depend largely upon the length of time employed in the treatment. This trouble, which is inherent in the method, does not often occur to a very appreciable extent, and the results obtained by different operators ordinarily agree fairly well.

6. **Barium in the Insoluble Residue.**—If an ore contains barium sulphate the latter will of course be found with the insoluble residue. Any soluble barium mineral in the presence of sulphates or sulphides will also be either wholly or partly changed to sulphate which will remain behind. When the insoluble residue is thus contaminated with barium sulphate, it is customary to determine the same and deduct it. This may be done as follows:

Fuse the mixed residue, after weighing, with about 3 grams of sodium carbonate, or mixed sodium and potassium carbonates, best in a platinum dish. Prolonged fusion is not necessary, as the object is simply to convert the barium sulphate to carbonate. After cooling, heat the fused mass with water until disintegration is complete and then filter through a 9-cm. filter and wash thoroughly to remove sulphates. The barium is left on the filter as carbonate, probably more or less impure. A little barium carbonate may also remain adhering to the platinum dish. Dissolve the latter in 5 cc. of strong hydrochloric acid diluted with about 10 cc. of water, and, if the amount of carbonate on the filter is small, dissolve it by pouring the acid mixture (best warmed) over it from the dish, receiving the filtrate in a large beaker. If the quantity of carbonate is large, rinse it from the filter to a small beaker and dissolve it by adding the acid in the dish, keeping the beaker covered with a watch-glass to prevent loss by spattering. Warm the solution and pour it through the filter to dissolve any carbonate remaining thereon. Wash the filter well with hot water. Dilute the filtrate to about 300 cc. with hot water, heat to boiling and precipitate the barium as sulphate, in the usual way (consult VII, 1), with dilute sulphuric acid. After standing, hot, for at least several hours, filter off the barium sulphate, wash with hot water, ignite, and weigh. The weight found, deducted from that of the original mixture, will leave the weight of the insoluble residue required.

7. Treatment of Ores and Slags that Gelatinize with Acids.—Some ores and furnace products are more or less completely decomposed by acids with the formation of gelatinous silica. Slag that has been suddenly chilled by dipping an iron bar into the molten mass and then plunging it with the adhering slag into cold water, may usually be entirely decomposed by acids. Roasted ores generally give gelatinous silica. If such substances are treated in the usual way, the gelatinous silica is liable to form a cake which adheres to the bottom of the flask and greatly hinders decomposition by surrounding particles of the ore with a more or less impervious coating. Material of this class may be treated as follows:

Moisten the 0.5 gram of substance in the flask (a 4-inch porcelain casserole or dish is better) with 2 or 3 cc. of water, and then add gradually about 10 cc. of strong hydrochloric acid, stirring or shaking the mixture at the same time to prevent coagulation. Cover the dish with a watch-glass and heat very gently with frequent stirring, until decomposition is as complete as possible, adding a little nitric acid if undecomposed sulphides still remain. Finally, remove and rinse off the cover, evaporate to dryness and finish in the usual way (3).

In the case of chilled slag the silica thus obtained is fairly pure. When there is much gelatinous silica especial pains should be taken in the heating after evaporation so as to dehydrate it as thoroughly as possible. In technical work it usually suffices to heat the dry residue for half an hour or an hour at a temperature of about 150° C. (cf. 11).

8. "True Silica," or "Silica by Fusion." *—These terms, in technical work, refer simply to the silica obtained by technical methods when the decomposition of the substance includes a fusion with alkali carbonate. The result is ordinarily much more accurate as regards actual silica than that obtained by the usual acid treatment, but it is not customary to follow all the requirements of exact analysis. In the case of ores containing considerable matter soluble in acids, or constituents undesirable in the fusion, such as sulphides and compounds of reducible metals, it is usually best to give a preliminary treatment with acids and confine the fusion to the residue finally obtained, which may consist of mixed silica and undecomposed matter. Such ore

---

* Insoluble silicates, such as clays, etc., are usually decomposed by this method. In accurate analyses the procedure may be according to **20**.

is decomposed and treated, according to its nature, precisely as described for INSOLUBLE RESIDUE (2) and the residue, containing the total silica of the substance, ignited in a platinum dish or crucible to burn off the filter-paper. It need not be weighed.

9. Mix the ignited "insoluble residue," or 0.5 gram of the finely pulverized ore or silicate, if no preliminary treatment was given, in a platinum crucible or small platinum dish with about 5 grams of sodium carbonate, or a mixture of 3 parts of sodium carbonate and 2 parts of potassium carbonate, which fuses more easily. If the ore, not subjected to previous acid treatment, contains, or is liable to contain, a small amount of sulphides, mix in a small pinch of potassium nitrate to insure their oxidation. Heat the mixture over a blast-lamp, very slowly at first. so as to expel moisture without mechanical loss of substance. Raise the heat very gradually, so that much of the carbon dioxide may be driven off without causing any spattering, before the mass has actually fused to a liquid. Finally heat strongly to complete fusion of the mixture and continue the heating until bubbling has practically ceased and quiet fusion is attained, In some cases this takes from one-half to one hour. If a crucible is used it is best to keep it covered to prevent loss from spattering; with a dish there is less danger. Cool the crucible or dish and digest the fused mass with hot water. As a crucible is too small to contain sufficient water, it is best to place it in a 6-inch porcelain dish, first detaching the cake, if possible, as follows: The crucible having been chilled quickly by dipping the bottom in cold water, a minute crack will usually develop between the sides of the cake and the crucible. Allow a drop or two of water to fall into this crack and then cautiously heat the crucible over a lamp. Under these conditions the cake will usually detach itself with a slight "pop." (Guiterman's method.) Place the crucible and cake in the evaporating-dish and warm with water until disin-

tegration is as complete as possible. Now remove and rinse off the crucible and cover. If not perfectly clean inside, dissolve the adhering residue with a little hydrochloric acid and add the solution to the main portion in the dish after the latter has been acidified. If a platinum dish has been used instead of a crucible, warming in the dish with one or two portions of water will usually suffice for disintegration and transfer to the porcelain dish. The last traces of residue may be removed by the acid used for acidifying.

10. Cover the evaporating dish and cautiously add hydrochloric acid in excess. It is not a good plan to add the acid before removing the platinum, since when manganese is present chlorine may be evolved and attack the platinum if the latter is exposed for any length of time. Nitric acid can be used and the danger avoided, but it is not so desirable a solvent.

Warm the mixture in the dish, and when the effervescence has sufficiently subsided remove and rinse off the cover and then support it on a glass triangle placed on the dish so as to allow a free circulation of air over the liquid. If the fusion has been successful and the decomposition is complete, practically everything should dissolve in the acid liquid and no hard, gritty, or dark particles should remain in the bottom of the dish.

11. The solution must now be evaporated to dryness. This may be done on a water-bath or as otherwise convenient,* except that care must be taken not to overheat at the end. When the mass is dry, remove the triangle and place it on top of the watch-glass and the latter directly on the dish. This is to prevent loss by any subsequent decrepitation. Heat now over a radiator or in an air-bath at a temperature of about 150° C. for 30 minutes

* With a water-bath the operation is sometimes very tedious. I usually support the dish on a scorifier or triangle placed in an iron sand-bath or radiator and have a low flame under the latter.

or more. The residue should finally appear perfectly dry and powdery under a glass rod. It is impossible to render the silica perfectly insoluble, so as to remove it pure at one operation, by any modification of the heating, but for ordinary technical results the above directions will suffice. Too high a temperature will cause a recombination of some of the silica with the bases present, and then, on the addition of acid, some of this silica will again be set free in a soluble form.

12. To the dried mass add about 20 cc. of strong hydrochloric acid and allow to stand, covered, at the ordinary temperature or slightly warmed, for about 10 minutes; then add about 100 cc. of hot water and heat to boiling, first removing and rinsing off the cover and triangle. After boiling a moment, allow the silica to settle and decant the clear liquid through a filter. It is best to wash the silica once or twice by decantation with hot water and then transfer it to the filter and finish the washing. If the upper edge of the filter appears brown, from the separation of a basic ferric compound, due to hydrolysis of ferric chloride, wash with warm dilute hydrochloric acid until clean.

Place the moist filter and contents in a weighed platinum crucible and ignite until perfectly white. Cool in a desiccator and weigh.

13. Testing the Weighed Silica.—When it is desired to test the purity of the silica obtained, it may be done as follows: Cover the silica in the platinum crucible or dish with 2 or 3 cc. of water to prevent violent action, and then add 2 or 3 drops of strong sulphuric acid and 3 to 5 cc. of pure hydrofluoric acid. Evaporate the mixture as far as possible on a water-bath, and then heat cautiously over a small free flame to complete dryness. If there is a residue add a little more hydrofluoric acid and again evaporate. Repeat this procedure until no residue remains, or there is no further diminution of the residue obtained. In the latter case add a drop or two of strong sulphuric acid, heat to dryness and ignite strongly over the blast-lamp, cool, and weigh. Unless the amount of residue thus found is large it will usually be sufficient to deduct its weight from that of the impure silica to obtain the true silica.

Barium sulphate in an ore would naturally contaminate the silica. Its amount could be determined and deducted as just described, or the weighed impure silica could be fused with alkali carbonate and the barium determined as sulphate, as described in 6.*

14. Determination of Silica in Ores Containing Fluorine.—Substances containing fluorine will sustain a loss of silicon by volatilization as fluoride if the solution from the carbonate fusion is subjected to the usual evaporation with acid. The following methods are applicable in such cases.

15. Seeman's Method.† — Fuse with alkali carbonate as directed in 9. Extract the melt with hot water and filter from the insoluble residue. Any barium present will remain with the residue as carbonate. Make the filtrate faintly acid with hydrochloric acid and then add an excess of a solution of mercury ammonium carbonate. Evaporate the mixture to dryness, add water to the residue and again evaporate to dryness. Take up the residue in hot water. It is easily filtered and washed and may be directly ignited and weighed to obtain the anhydrous silica.

The mercury-ammonium carbonate may be prepared by adding to a quantity of mercuric oxide a sufficient excess of a

* A more common method at western smelting-works is to filter the aqueous solution of the melt of the original fusion before acidifying. The barium is thus practically all removed as carbonate at this point.

† F. Seeman, Zeit. f. anal. Chem., LXIV, 343–387.

cold saturated solution of ammonium carbonate to form a clear solution.

16. **Method of Berzelius.**—Thoroughly extract the melt from the alkali carbonate fusion (9) with hot water and filter from the insoluble residue. (Any barium will remain behind as carbonate.) To the filtrate add about 4 grams of solid ammonium carbonate, heat for some time at about 40° C. and then allow to stand over night. Filter off the precipitate and wash with water containing ammonium carbonate. Reserve the silica thus obtained.

Some silica still remains in the filtrate. To recover this, add to the solution 1 or 2 cc. of ammoniacal zinc oxide solution (made by dissolving moist zinc oxide in ammonia water), boil until the ammonia is all expelled and then filter off the precipitate of zinc silicate and oxide and wash with water. Spread out the filter on a watch-glass, and, with a jet from the wash-bottle, rinse as much of the precipitate as possible into an evaporating-dish. Pour a little dilute hydrochloric acid over the filter and any remaining residue, letting it run into the dish, then burn the filter and add the ash. Add a little more hydrochloric acid to the contents of the dish, if necessary to decompose the precipitate, and then evaporate on the water-bath and separate the silica as usual (11). Filter off this silica and reserve it.

The insoluble part of the melt may contain silica. Dissolve it in hydrochloric acid and separate the silica by evaporation as usual. (Any barium will pass into the filtrate as chloride.)

Finally ignite and weigh all three silica precipitates together.

17. **Decomposition of Silicates by the Lead Oxide Method of Jannasch.**—This method is applicable in the case of pure silicates and admits of determination of the alkalies in the same portion. Pure lead oxide is required, but as the commercial article is not safe, it is best to use the carbonate, which is con-

verted into the oxide by ignition. It may be prepared as follows: To a boiling solution of lead acetate add the theoretical amount of ammonium carbonate. Wash the precipitated lead carbonate several times by decantation with hot water, then transfer it to a hardened filter (to avoid contamination with loosened fibers), wash completely and allow to drain thoroughly, or better, use suction. Carefully remove from the filter and dry completely in a porcelain dish on the water-bath.

18. Mix 1 gram of the very finely powdered substance with 10–12 grams of lead carbonate. Place the carbonate in the crucible first, then add the weighed substance and mix very thoroughly with a platinum spatula. Cover the crucible and heat gently, below the fusing-point of the mixture, for 15 or 20 minutes, to expel most of the carbon dioxide. Now heat more strongly, being careful to have a non-luminous oxidizing flame, until the fusion is complete and keep it at this temperature for 10 or 15 minutes. Only the lower third of the crucible should be heated to redness. Cool the crucible quickly by dipping the bottom in hot water, then place it, together with the cover, in a platinum dish and cover with hot water. After adding sufficient strong nitric acid to insure solution of the lead, heat on a water-bath until disintegration is complete. Remove the crucible and cover when the adhering portions of the melt have dissolved or separated therefrom. To assist in the disintegration stir the mixture frequently and break up hard lumps as much as possible. Only slightly colored flocks of silicic acid should finally remain floating in the liquid. Now evaporate to complete dryness. Moisten the dry mass with strong nitric acid and again evaporate to dryness. Add 20 cc. of strong nitric acid to the dry residue, allow to stand 15 minutes and then add 100 cc. of water. Heat the mixture for 20 minutes on the water-bath and then filter. Wash the silicic acid first with hot water acidulated with nitric acid, and then with pure hot water. Ignite and weigh in the usual manner.

19. **Common Errors in the Determination of Silica.**—Under this heading Dr. Hillebrand has published an exhaustive paper in Vol. XXIV of the Journal of The American Society. While the technical chemist, working by conventional methods, might regard his findings as of minor importance, they should nevertheless always be borne in mind. They are as follows:

1. Silica cannot be rendered wholly insoluble by any number of evaporations with hydrochloric acid when followed by a single filtration, no matter what temperature may be employed. Two or more evaporations alternating with filtrations are necessary to secure satisfactory results.*

2. The generally accepted view that silica passing into the filtrate is wholly thrown down by ammonia or sodium acetate in presence of much alumina or iron is incorrect.

3. Silica is appreciably soluble in melted potassium pyrosulphate, and consequently, when silicious oxides of iron and aluminum obtained in analysis are thus fused, their silica contents are only in small part left undissolved when the fused mass is taken up with water or acid. These last two sources of error may be avoided by separating all silica at the start as above.

4. The need of long blast ignition of the silica in order to get the correct weight is proved; continual loss of weight being invariably observed for 30 minutes or more.

20. **Accurate Method for Silica.**—Treadwell,† for an accurate silica determination, directs as follows: Evaporate the

---

* Hillebrand (private communication) says: "Where I have fused a gram of rock (silicate) with 4–6 grams of sodium carbonate, I always recover about 3 milligrams of silica from the iron and alumina, even after two evaporations and filtrations."

† Anal. Chemistry, Hall, 2d Ed., Vol. II, p. 445.

dilute hydrochloric acid solution of the residue after the fusion, or the original material if entirely decomposed by acid, in a porcelain dish, to dryness on the water-bath.* Stir frequently until the residue is obtained as a dry powder. Moisten this with strong hydrochloric acid, cover the dish and allow to stand at least 20 minutes at the ordinary temperature, to insure the changing to chlorides of any basic salts or oxides formed during the evaporation. Then add 100 cc. of water, heat to boiling, allow the silicic acid to settle and decant the clear liquid through a filter supported upon a platinum cone in a funnel. Wash the residue three or four times by decantation with hot water, then transfer to the filter and wash with hot water until free from chloride. If the silicic acid is not perfectly white, but brownish from basic ferric salt, drop strong hydrochloric acid around the upper edge of the filter and immediately wash it down with a stream of hot water. Repeat this until the filtrate comes through perfectly colorless. Finally, dry the precipitate by suction, place it in a platinum crucible and set it aside for the time being. The separation of the silica is now by no means quantitative; as much as 5 per cent. of the total amount may remain in the filtrate. To remove this, evaporate the solution once more on the water-bath to dryness, keep at this temperature for 1 or 2 hours (or more), moisten with a few cubic centimeters of concentrated hydrochloric acid and allow to stand not more than 15 minutes. Too long a contact with the acid at this point will cause some of the silicic acid to go into solution. Add hot water, filter the residue through a new, and correspondingly small, filter and wash with hot water.

The amount of silicic acid now remaining in the filtrate should

---

* This is sometimes a very tedious operation. Hillebrand says of the drying: "It can safely be done at temperatures higher than that of the water-bath." — *Private communication.*

not exceed 0.15 per cent. of the total quantity and may usually be neglected. It can be removed, however, by a third evaporation to dryness.

The filters containing the silica are ignited wet in a platinum crucible, and finally over the blast-lamp, and weighed.

Treadwell further remarks: In order to make the separation of silicic acid quantitative it has been proposed to heat the residue obtained by evaporation at 110°–120° C. It has been found, however, that nothing is gained by this practice, as some silicic acid is obtained in the filtrate and the deposited silicic acid is less pure than when dried on the water-bath. When magnesia was present more silicic acid was found in the filtrate after igniting at 280° than when dried on the water-bath. This is due to the fact that magnesia formed by hydrolysis reunites with the silica to form magnesium silicate, and the latter is decomposed by hydrochloric acid with the formation of soluble silicic acid. It is, therefore, not advisable to attempt to dehydrate the silica at a temperature higher than that of the water-bath.

21. Methods in Use at the Colorado Plants of the American Smelting and Refining Co. for the Determination of Insoluble Matter in Ores, etc.*—The methods used in the different laboratories may differ somewhat in unimportant details, but in general are as follows:

Unless there is some special reason otherwise, one-half gram is taken in all cases for the determination.

Sulphides.

Treat in No. 0 beaker, or small casserole, with strong nitric acid, 7–10 cc. Heat gently until strong action has ceased. Evaporate to dryness and bake until free from acid. Cool, add

* Western Chemist and Metallurgist, III, 120.

about 20 cc. hydrochloric acid (1 part acid and 1 part water) and heat until solution is as complete as possible. Filter, wash first with hot dilute hydrochloric acid, then hot water. Ignite and weigh.

Occasionally some ores high in zinc or lead sulphide may be treated to advantage, first with hydrochloric acid, and then following as above, but the result should be the same in either case if properly carried out.

Oxidized Ores in General.

Treat with strong hydrochloric acid, 7–10 cc. Boil until dissolved; add a little nitric acid (0.5 cc. are usually plenty); evaporate to dryness and bake to the complete expulsion of acid fumes. Cool; take up with hydrochloric acid and proceed as with sulphides.

With some ores containing carbonaceous matter, it may be necessary to bake for a long time, as such residues often hold acid very tenaciously.

Manganese Oxides.

Nitric acid is omitted entirely with these.

Leady Oxide Ores.

Some of these usually yield gelatinous silica and must be carefully dehydrated at not too high a temperature.

Oxidized Material Which has been Strongly Ignited.

Digest with hydrochloric acid without boiling at first, then evaporate nearly to dryness, add a few drops of nitric acid and evaporate to complete dryness. Occasionally it may be necessary to repeat this treatment, which is a matter of individual judgment. Finally bake to complete expulsion of acid and proceed as before.

Roasted Ores, Acid Works' Residues, Etc.

Digest with hydrochloric acid, without boiling, until oxidized portion is dissolved, add 2–3 cc. of nitric acid, boil to decompose sulphides, evaporate to dryness and proceed as before.

Ores or Products Containing Magnetite

Some mixed oxidized and sulphide ores, and also often furnace mattes, contain magnetite. When this is known or suspected, the material is treated as above under "Roasted Ores, etc.," except that nitric acid in larger quantity may sometimes be necessary.

Barium Sulphate Ores.

Treat with 10 cc. hydrochloric acid (1 part acid and 1 part water), boiling a few minutes, add 4–5 cc. nitric acid, and after action has ceased evaporate to dryness. Bake and proceed as usual.

After total insoluble is weighed, fuse with sodium carbonate or mixed carbonates, digest the melt with water until disintegrated, filter and wash. Wash out the fusion crucible with 3–5 cc. hydrochloric acid (1 : 1) and with this dissolve the residue on the filter, being careful that the residue is all dissolved and the filter washed clean. Precipitate the barium in boiling solution with a slight excess of dilute sulphuric acid, and allow to stand at least 2 hours before filtering. The barium sulphate so obtained is deducted from the total insoluble to give the insoluble residue.

General Notes.

Care must be taken to insure the removal of all lead salts from the insoluble residue. Filtrations should be conducted rapidly and with hot solutions, in which case washing with hot dilute hydrochloric acid is perfectly safe. However, washing with hot ammonium chloride or acetate solution is often used and has no objections. This removal of lead needs special care in barium sulphate ores, as this latter seems to render the complete solution of the lead salts more difficult.

The use of the Baker and Adamson No. A filter paper is recommended as less likely than most papers to allow fine precipitates to pass through.

CHAPTER XXV.

SULPHUR.

(See Appendix.)

1. Method for Ores, etc., Giving Total Sulphur.*—Prepare an intimate mixture of 1 part of dry sodium carbonate and 4 parts of zinc oxide. Weigh 0.5 gram of the ore into a small platinum or porcelain dish, add 3 grams of the above mixture and mix thoroughly, best by rubbing together with a small agate pestle. Cover with 2 grams more of the zinc oxide mixture.† If a platinum dish is used, set it in a hole in a piece of asbestos-board (to exclude sulphur coming from the gas), cover partially with a piece of thin asbestos-board, and heat over a Bunsen burner to bright redness for about twenty minutes. If a porcelain dish is used it is difficult to attain the proper heat with a Bunsen burner, and it is best to heat it in a muffle or over a suitably adjusted blast-lamp. Allow to cool and transfer the mass to a 250-cc. beaker. Rinse out the dish thoroughly with hot water and make up the bulk in the beaker to about 50 cc. Heat the mixture to boiling, stirring well, and then filter through an 11-cm. filter and wash the residue at least ten times with hot water. Receive the filtrate in a 600-cc. Erlenmeyer flask. Add a drop of phenolphthalein solution as indicator, make just acid with strong hydrochloric acid, and then add 4 cc. in excess. Dilute the solution to about 400 cc. with hot water, heat to boiling and precipitate the sulphur with barium chloride, as described in 6. For ordinary technical work I do not follow all of Folin's direc-

---

* The use of the zinc oxide mixture is described by Ebaugh and Sprague, Jour. Am. Chem. Soc., XXIX, 1475.

†See Note, page 245.

tions. Continue as follows: After adding the barium chloride, as described, to the boiling solution, allow the mixture to stand, hot, until the liquid above the precipitate has become perfectly clear—perhaps an hour. Filter through a double 11-cm. filter. No appreciable amount of barium sulphate should run through. Unless, however, the filtrate appears practically clear, always filter a second time, which will usually suffice. Wash the precipitate ten times with hot water. Transfer the moist filter and precipitate to a clean smooth "annealing-cup" and ignite, with free access of air, over a Bunsen burner, or in a muffle, at a gentle heat. A high heat, such as that of a blast-lamp, is neither necessary nor desirable. The ignited barium sulphate should be perfectly white. Cool in desiccator, transfer to the scale-pan by tapping and brushing with a camel's-hair brush, and weigh. Multiply the weight of the barium sulphate by 0.1373 to obtain the weight of the sulphur.

A platinum or porcelain crucible may, of course, be used for the ignition of the barium sulphate instead of an annealing-cup.

It is best to run a blank, once for all, with all the reagents employed, and always deduct for any sulphur thus found.

It is usually unnecessary to test the filtrate from the barium sulphate with more barium chloride, unless the precipitate is large in amount, or the percentage of sulphur found approximates the total amount that 10 cc. of the barium chloride solution are capable of precipitating (6).

2. Acid Method for Ores (Not Giving the Sulphur of Barium Sulphate).—Treat 0.5 gram of the ore in an 8-oz. flask with 10 cc. of strong nitric acid. Heat very gently until the red fumes have somewhat abated, and then add potassium chlorate in small portions at a time (say 0.2–0.3 gram), until any free sulphur that has separated is entirely oxidized and dissolved. The acid should not be boiled violently, as this would unnecessarily weaken it. On the other hand, it is best not to allow it simply timmer, as the explosive gases from the decomposing chlorate may then collect in the flask and produce annoying, although not dangerous, explosions. When the sulphur has entirely disappeared the solution should be boiled to complete dryness. This operation may be hastened by manipulating the flask over a free flame. After cooling, add 5 cc. of strong hydrochloric acid. This should be done cautiously to avoid a too violent reaction with the undecomposed potassium chlorate that may be present. If iron oxide, etc., still remains undissolved, gently heat the hydrochloric acid mixture until solution is as complete as possible, adding more acid if necessary. Finally, boil to dryness, then add 5 cc. more of the hydrochloric acid and again boil to dryness. This is to decompose nitrates and expel all nitric acid. Take up once more in 5 cc. of hydrochloric acid and dilute with about 100 cc. of cold water.* Make alkaline with ammonia and then add 10 cc. of a saturated solution of ammonium carbonate. This latter is to convert any lead sulphate to carbonate and thus render the combined SO_3 soluble, as ammonium sulphate. Heat to boiling, allow the ferric hydroxide, etc., to settle, and then filter and wash very thoroughly with hot water, receiving the filtrate in a 600-cc. Erlenmeyer flask. Proceed with the filtrate as described in 1.

3. Modification of the Acid Method for Ores Containing Barium Sulphate.—As barium sulphate remains practically unaffected by the above acid treatment, any sulphur thus combined in an ore will require another method for its solution. When the total sulphur is required in ores containing barium sulphate, the procedure may either be according to 1, or the acid method may be modified as follows: Begin the analysis according to 2, and proceed as described until all nitric acid has been

* If the solution is hot when made alkaline with ammonia, some basic ferric sulphate is liable to separate.

expelled and the residue has been taken up for the last time in 5 cc. of hydrochloric acid. Now dilute with about 30 cc. of water, add 5 grams of pure ammonium chloride and heat to boiling. The ammonium chloride is to hold any lead in solution. Filter and wash with hot water. Reserve the filtrate, which contains all the sulphur that yielded to the acid treatment. Ignite the insoluble residue, containing the barium sulphate, in a platinum dish or crucible to burn off the filter-paper. When cool, add a little sodium carbonate, or mixed sodium and potassium carbonates, and fuse to decompose the barium sulphate and form soluble alkali sulphate. A prolonged fusion is not necessary. Cool, take up the mass in hot water, warming until thoroughly disintegrated, and then filter and wash with hot water. Make the filtrate slightly acid with hydrochloric acid, keeping the beaker covered to avoid loss by spattering, and then add to it the reserved filtrate. The united filtrates are now cooled (cf. foot-note, p. 240), made alkaline with ammonia, ammonium carbonate added, and the determination proceeded with in the usual way.

4. Waring's Method for Sulphur in Ores, etc.*—To 1 part of the ore, ground impalpably fine, add 1 part of potassium chlorate and 2 parts of sodium carbonate. Mix and grind thoroughly together in an agate mortar. Add 4 to 6 parts of pure precipitated manganese dioxide and again grind together. Prepare a platinum or porcelain crucible to receive the mixture by lining its sides and bottom with magnesium oxide with the aid of the agate pestle (upper end). Pour and brush the mixture into the cavity and cover with a little additional manganese dioxide. Heat the crucible very gradually to a moderate red heat (750° or 800° C.), and maintain this temperature for 15 or

* Communicated by W. Geo. Waring.

20 minutes. When cold, invert the crucible over a casserole or beaker so the calcined but yet pulverulent mass may drop out. Add 40 or 50 cc. of hot water, boil thoroughly, allow to settle and decant through a filter. Repeat the washing by decantation three or four times and finally transfer the residue to the filter and wash with hot water. A few drops of bromine water may be added with the last washing. Slightly acidify the filtrate with hydrochloric acid, heat to boiling and precipitate with hot barium chloride solution.

The essential feature is the intimate mixture by grinding. If this is not thoroughly done, a little sulphur may sometimes be found in the residue upon treatment with hydrochloric acid, etc. The residue should always be tested in this way, although Waring states that he has never found it to contain sulphur except when the grinding was omitted and the reagents and mineral mixed merely with a spatula, or, in some cases, when potassium nitrate was used instead of chlorate. Generally, the nitrate may be used successfully. The reagents should be tested by a blank and the sulphur found deducted. The magnesium oxide serves two purposes,—preventing adhesion of particles of the assay to the crucible, and also preventing silicic acid from going into the aqueous solution and producing erroneous results.* Waring states that no other method he has tried yields such accurate results.

5. Method for Roasted Ores, and Ores Containing Much Copper.—By the wet method, some of the sulphur of roasted ores is liable to escape as hydrogen sulphide. Also, if the solution to which ammonia is added contains much copper, some of the latter is liable to separate as a basic sulphate, and, furthermore, the copper in the filtrate will contaminate the barium sulphate.

* Hillebrand states that silicic acid in dilute solution is not precipitated with the barium sulphate. Bulletin 176 of the U. S. Geological Survey (1900), p. 106.

Both of these sources of error may be avoided as follows: Fuse 0.5 gram of the ore in a large porcelain crucible with 25 grams of a mixture of 6 parts sodium carbonate and 1 part potassium chlorate (Böckmann's method). Contamination with sulphur of illuminating-gas is best avoided by placing the crucible within a hole in a piece of asbestos board. Heat gently at first, and finally until the evolution of oxygen ceases. Cool, extract with water and filter, washing thoroughly with hot water. Acidify the filtrate with hydrochloric acid and continue as described in 1.

6. Notes on the Precipitation of Barium Sulphate.—According to a series of most careful tests by Otto Folin,* most sulphate precipitations, carried out by the usual methods, are liable to considerable errors, which may be either losses or gains. Folin shows that accurate results may, in ordinary cases, be obtained as follows:

Have the barium chloride solution of 10 per cent. strength. Only a slight excess is necessary, and it should be allowed to drop through a capillary opening at the bottom of a small funnel, so that 10 c.c. will run through in from 2 to 10 minutes.

The precipitation is most conveniently made in an Erlenmeyer flask, with the dropping-funnel placed in the neck. The sulphate solution should be hot and contain an excess of from 1 to 4 c.c. of strong hydrochloric acid per 150 c.c. of solution.† The sulphates present should not yield over 0.200 gram of barium sulphate per 100 c.c. After the addition of the barium chloride, allow to stand until cold and then filter, washing with cold water. It is best to use a porcelain Gooch crucible for the filtration, as when paper is used there is an appreciable mechanical loss in the subsequent ignition. Even with a Gooch crucible there must be certain precautions in the ignition to avoid

* Jour. of Biological Chemistry, 1, 131.

† An excess of 0.3 cc. per 100 cc. of solution is the best acidity.

loss. Have the crucible covered and protect the bottom by standing the crucible on the cover of a platinum crucible placed on a triangle. Ten minutes will suffice for the ignition in the absence of organic matter.

If a paper filter is used, mechanical loss will occur during the ignition, in spite of every precaution. The loss may be reduced to a minimum by regulating the heat at the beginning so that the filter will char slowly without taking fire. This method will suffice for ordinary technical work.

10 cc. of a 10 per cent. solution of barium chloride will precipitate 0.1416 gram of sulphur, or, on the basis of 0.5 gram of ore, 28.32 per cent. On the same basis the precipitating solution should not contain more than 5.6 per cent. of sulphur per 100 cc. For an ore containing 20 per cent. of sulphur the solution should be diluted to about 400 cc. and have an excess of 1.2 cc. of strong hydrochloric acid. When more than this amount of sulphur is possibly present, it is better to take less than 0.5 gram of the ore rather than largely increase the volume of the solution for precipitation.

7. Determination of Sulphur in Liquid Fuel.*—Treat 2 or 3 grams of the fluid with 4 cc. of fuming nitric acid in a large platinum crucible covered with a watch-glass. When, after several hours, the reaction in the cold has ceased, heat the crucible on a gently-warmed water-bath. When the mass has quieted down remove the cover and continue the heating until the material is dry. Now add 6–8 grams of a mixture of calcined sodium carbonate (free from silica and sulphate) with 2 parts of potassium nitrate and heat over a rosette-burner, stirring with a platinum rod as soon as the mass softens. Cover with more of the nitrate mixture and continue the heating until the com-

* A. Goetzl. Zeit. f. angew. Chem., 1905, No. 30, p. 1528.

bustion is complete. Allow to cool, dissolve the melt in hot water and transfer to a beaker. Make slightly acid with hydrochloric acid, dilute sufficiently, heat to boiling and precipitate with barium chloride in the usual manner, finally igniting and weighing the $BaSO_4$.

Note to Paragraph 1.—There should always be present at least twice as much sodium carbonate as would be required to combine with the sulphur, arsenic, etc., in the ore.

CHAPTER XXVI.

TIN.

THE following is the best method with which I am acquainted for the technical determination of tin in ores, etc. It is very rapid (usually requiring less than an hour) and appears to answer every ordinary requirement. According to Parry * it is not as accurate for high-grade material as his reduction method, described below (10).

1. **Author's Modification of Pearce's Method (6).**—In a thin spun-iron crucible of about 60 cc. capacity, melt about 8 grams of sodium hydroxide. (I take about 3 inches of the stick hydroxide, broken into short pieces.) Heat until the moisture is expelled and quiet fusion attained. After cooling, add 0.5 gram of the finely ground ore, cover the crucible with a loosely-fitting porcelain cover, and heat, at first very cautiously to avoid spattering, and then with the full flame of a Bunsen burner until the fusion is quiescent. Remove the cover and pour the melt into a clean metallic dish floating in a beaker of water. I use a 2½-inch nickel dish. It is best to cover the hot cake with a small porcelain crucible cover, dropping within the dish, to prevent mechanical loss in case the cake cracks and flies apart violently. Place a little cold water in a 5½-inch casserole and set the hot crucible therein; then turn the latter on its side so as to admit the water, and heat to boiling. Move the crucible about with a glass rod,

* London Mining Journal, Sept. 25, 1909. See also Parry's excellent book, 'The Assay of Tin and Antimony."

and when the outside is clean wash it off so as to remove the crucible with the fingers. The inside of the crucible may still contain some of the undissolved melt. Pour in a little water, acidify with hydrochloric acid, and bring the dilute acid, by means of a glass rod, in contact with any adhering melt. When all is dissolved, wash the solution into the casserole. Now add the detached cake and the crucible-cover with the top (provided with a loop) up. Cover the casserole and add about 30 cc. of strong hydrochloric acid, which should prove an ample excess. Heat to boiling and the cake will quickly dissolve. Now remove and wash off the watch-glass covering the casserole, and, by means of a bent iron wire, lift out the crucible-cover and wash it.

2. Transfer the solution, which should contain no residue of undecomposed ore (although there may be some scales of iron oxide from the crucible) to a tall narrow beaker, or, better, a 12 or 16-oz. conical flask with a wide mouth. The flask should be marked at the 200 cc. point. Add to the solution 50 cc. of strong hydrochloric acid and then dilute with hot water to 200 cc. Prepare a piece of sheet nickel by rolling a strip about 7 inches long and 1 inch wide into a loose coil. This may be introduced into and removed from the flask by means of a nickel wire or strip, either permanently attached to the coil or provided with a hook at the end. Place the coil in the flask, cover the latter and heat the liquid to boiling. Maintain in gentle ebullition for 20 minutes, when the tin in solution will all have been reduced to the stannous condition. Remove from the heat and at once add a small piece of marble (equivalent to about $\frac{3}{8}$-inch cube) so as to fill the flask with carbon dioxide, and place the covered flask in running water to cool as rapidly as possible. When the liquid is at room temperature, remove the coil of nickel (which will not be greatly attacked), rinsing it

off with a mixture of 1 part of strong hydrochloric acid and 3 parts water, so as not to reduce the acid strength of the solution. Now add a little starch liquor (XIII, 2) and titrate at once with standard iodine solution to a permanent blue color. The carbon dioxide in the flask is usually quite sufficient as a preventive of oxidation, but it is well to agitate the mixture as little as possible during the titration and to deliver the iodine solution with the tip of the burette well down in the flask.

None of the ordinary constituents of ores interfere with this method. Both arsenic and antimony in the *ous* condition in weak acid solution consume iodine, but in a solution containing at least one-fourth of its volume of strong hydrochloric acid they are entirely without effect.

3. Instead of reducing the tin solution as described in 2, the following procedure may be adopted to insure accuracy:* Transfer the solution from the casserole to a flask of about 500 cc. capacity and marked at the 200 cc. point. Add 50 cc. of strong hydrochloric acid and dilute the solution with hot water to 200 cc. Add about 1 gram of finely powdered C. P. antimony and boil the mixture gently for fifteen minutes, shaking occasionally. Now discontinue the heat and connect with an apparatus capable of giving a rapid current of carbon dioxide. The connection is made by means of a stopper carrying two tubes. The first, which is connected with the carbon dioxide apparatus, dips below the surface of the solution in the flask, and the second or outlet tube simply passes through the stopper and has the outer end bent downward so as to dip slightly below the surface of some water or mercury. This allows any tendency to back pressure to be instantly detected. While passing a rapid stream of gas,

* W. H. Low, Jour. Am. Chem. Soc., XXIX, 66; also Ibbotson and Brearly, Analyst, 1902, 25.

again heat the contents of the flask to boiling and boil for two or three minutes. Cool quickly by surrounding the flask with cold water, and take care that the current of carbon dioxide is strong enough to offset the back pressure due to sudden condensation. When cold, loosen the stopper somewhat and introduce about 5 cc. of starch solution, cork quickly and take to the burette. Introduce the spit of the burette well into the flask and titrate with the standard iodine solution, while rotating the flask gently and avoiding violent agitation.

4. Ores containing pyrites should be given a preliminary treatment with aqua regia. Filter off the insoluble residue containing the tin, ignite it at a low temperature in the iron crucible to be used for the fusion, and then proceed with assay as usual.

In giving this preliminary treatment to material containing tin in a soluble form, it is of course necessary to evaporate the aqua regia mixture to hard dryness and take up again in dilute nitric acid before filtering.

5. The iodine solution may be prepared by dissolving about 11 grams of iodine in a little water with the addition of about 20 grams of potassium iodide, and diluting to 1 liter. On the basis of 0.5 gram of ore taken for assay, 1 cc. will equal about one per cent. of tin. Standardize as follows: Weigh exactly 0.2 gram of pure powdered arsenious oxide, place in a 6-oz. flask and dissolve by warming with a little dilute sodium hydroxide solution. When dissolved, dilute to about 100 cc. with cold water, add a few drops of phenolphthalein solution as indicator and make slightly acid with hydrochloric acid. Now cool to room temperature or cooler. Finally add 3–4 grams of sodium acid carbonate and a little starch liquor * and titrate to a permanent blue color with the iodine solution. Pay no attention to a brownish discoloration toward the end, but proceed slowly

* If an acid starch solution is used add it before neutralizing.

until a single drop of the iodine produces a strong permanent blue color. 0.2 gram of arsenious oxide = 0.2404 gram of tin, from which the value of the iodine solution may be calculated. The solution does not change materially during a week, but it is not safe to use a standard that is older.

6. Method of E. V. Pearce for the Determination of Tin in Ores, etc.—Fuse * from 8 to 10 grams of stick sodium hydroxide in a $2\frac{1}{4}$-inch nickel dish with the addition of a sprinkling of finely-powdered charcoal, keeping the dish covered with a porcelain crucible-cover. When the sodium hydroxide is thoroughly fused, remove the lamp and allow to cool slightly.

7. Weigh from 0.2 to 0.5 gram, according to its supposed richness, of the finely powdered ore, place it in a thin layer over the cooled melt in the dish, and sprinkle a little more charcoal over the ore. Cover the dish and heat again, cautiously at first to avoid spattering, finally with the full power of a Bunsen burner for 5 or 10 minutes, or until no further action is seen. Allow to cool, and when cold the separation of the cake may be effected by squeezing the dish in the hand. Transfer the cake to an evaporating-dish (a glazed iron dish is ordinarily employed), add water, and thoroughly clean and wash the nickel dish and cover. Now cover the evaporating-dish and add sufficient strong hydrochloric acid to effect a complete solution of the cake, with the

---

* For this description I am indebted to Dr. Richard Pearce, who has kindly written it with the consent of Mr. E. V. Pearce, his nephew. Dr. Pearce describes the method as he has seen it in daily use at the works of Messrs. Williams, Harvey & Co., Ltd., in Cornwall, with which his nephew is connected. In regard to its accuracy, Dr. Pearce states: "I have seen the method tried on various ores and mixtures, the latter being prepared with percentages of tin known only to myself, and am quite satisfied with its accuracy and rapidity. In my opinion it is far ahead of any method which has yet been suggested for the assay of tin, especially in low grade ores."

exception, possibly, of some gelatinous silica. Under ordinary circumstances the latter may be disregarded, but, if desired, it may be removed by filtering through glass wool. Transfer the solution to a tall narrow beaker, add a few rods of iron (nail-rod),* about 3 or 4 inches long, cover the beaker and warm it gently to effect a free evolution of hydrogen. The ferric chloride is reduced to a ferrous salt, the yellow color giving place to a pale green, and then the stannic chloride is reduced to stannous chloride. The whole reduction is complete in 30 minutes or ever less. Now place the beaker and contents in a dish in which cold water is circulating until the solution is as cold as possible. Remove and wash the iron rods, add a little starch liquor, and the solution is then ready for titration with a standard iodine solution.

8. The latter is prepared in the usual way by dissolving iodine in a solution of potassium iodide, and is standardized with pure tin or pure arsenious oxide. About a N/10 strength is convenient.

9. In the case of ores containing pyrites, the latter should be oxidized before the fusion. To avoid possible loss in roasting, it is best to treat with aqua regia until the sulphides are decomposed, then dilute, filter, and ignite the residue, containing all the tin oxide, at a low temperature in a porcelain capsule (it being unnecessary to burn off all the carbon) and proceed with it as described above.

10. In regard to his method, Mr. E. V. Pearce writes: "I should explain to you that the merit of my process lies in the ease with which the tin is obtained in solution. The titration with iodine which I am using has been elaborated by L. Parry and is described in detail in his book, "The Assay of Tin and

---

* Mr. Pearce now uses sheet nickel as the reducing agent. I have likewise found nickel preferable to iron.

Antimony." The advantages of the process, in conjunction with the iodine titration, over any other known tin assay are:

"1. The ease with which all the tin is obtained in solution;

"2. The absence of any filtrations;

"3. No separation of other metals is necessary;

"4. The speed with which an assay can be made, — $1\frac{1}{2}$ hours being sufficient for any ores."

NOTE. — Mr. E. V. Pearce has recently called my attention to the fact that the omission of charcoal in the decomposition, according to my modification of his method, is apparently permitted by the use of an iron crucible. When a nickel crucible is used, the decomposition is frequently incomplete if charcoal is omitted, but iron filings appear to serve equally as well as charcoal. The iron of the crucible is therefore a factor in the reactions, and Mr. Pearce prefers the method I describe, as being more convenient than the use of either charcoal or iron filings.

11. **Parry's Method for the Assay of Tin Ore.***—Tin ores may be either pyritic or non-pyritic. Pyritic ores holding more than two per cent. sulphur must be, and non-pyritic ores may be, treated with nitric acid before reduction, as tin sulphide is volatile at a red heat. Weigh 5 grams of the ore, ground as finely as possible, either directly into a porcelain boat, or treat in an evaporating dish with a watch-glass cover with 20 cc. of dilute nitric acid, and carefully evaporate to complete dryness. Digest the residue with dilute nitric acid and filter. Dry the washed residue, ignite it in the dish, and then transfer it to a porcelain boat and heat it to low redness for two hours in a slow current of hydrogen or coal gas. Coal gas is much more convenient to use and quite as effective as hydrogen. The boat is $2\frac{1}{2}$ by $\frac{1}{2}$ inch, and two at a time may be placed in a porcelain or glass tube 12 inches long and $\frac{3}{4}$ inch bore, which is then placed in the reduction furnace. A very convenient form of gas reduction furnace, with clay body,

* L. Parry, London Mining Jour., Sept. 25, 1909; also Parry's "The Assay of Tin and Antimony."

brass gas jets, and asbestos rings to fit over the ends of the tube against the clay covers, is made in 6-inch lengths. The ends of the tube should project about 3 inches at each end of the furnace, and should be closed with rubber stoppers fitted with glass tubes as shown; the escaping gas (about 4 bubbles per second) is passed through dilute hydrochloric acid, in which any volatilized tin is condensed. In ores holding not more than 2 per cent. sulphur—that is to say, in 99 out of 100 samples of black tin or tin barilla, as bought, treatment with nitric acid may be omitted. In these direct reduction assays the tube is washed out, after the reduction is complete, with hydrochloric acid and potassium chlorate, also the glass leading tube. The solution is mixed

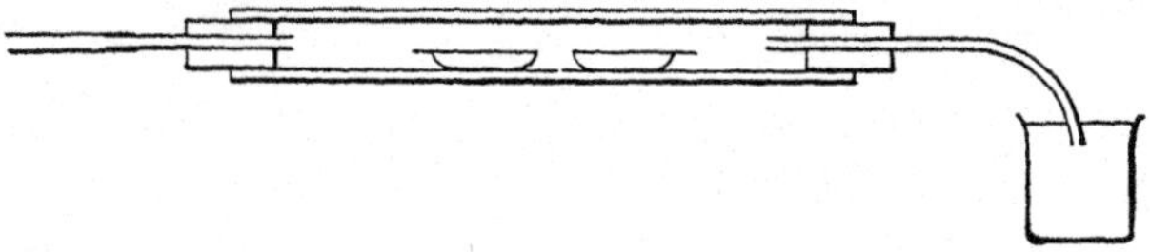

FIG. 19.

with that through which the gas bubbled, reduced with iron and titrated for tin with iodine. This generally adds 0.1 or 0.2 per cent. on the main tin assay. Where treatment of the ore with nitric acid is adopted as a method, the tube may still be similarly washed out, and, together with the gas liquor, tested for tin, as a check, for instance, against volatilization of tin on account of sulphur in the gas. A two-way gas branch is used; one jet supplies gas to the tube, the other supplies gas for heating the furnace. After the reduction allow the boats to cool in the furnace, and, when cold, transfer each boat and its contents to a 400-cc. beaker and treat with 150 cc. of hydrochloric acid and 5 cc. of nitric acid. Allow the assay to stand in a warm place until the action abates, then boil for a few minutes, dilute with an equal bulk of water and filter. Wash the residue well with hot acid water, then with hot water, and then dry and ignite it. Fuse the ignited residue with 4 or 5 parts of a mixture of sodium carbonate (free from chlorine) and borax, in a platinum crucible. Dissolve the melt in hydrochloric acid and precipitate with zinc, as usual. The residue rarely holds more than 0.5 per cent. of the total tin in the ore. Instead of fusing the residue as above, it may be fused with about 2 inches of stick sodium hydroxide in an iron crucible over a Bunsen burner. Dissolve the melt in hydrochloric acid and precipitate on zinc as usual. Make up the main solution to 500 cc., pipette off the equivalent of 1 gram into an 8-oz. flask, reduce with iron wire and filter. Precipitate the filtrate with strips of sheet zinc as usual. Dissolve the metallic sponge in hydrochloric acid and titrate at the boiling-point with standard ferric chloride solution.

In precipitating the tin with zinc, neutralize the solution (which should be freely acid) with thin strips of zinc. Use an excess of zinc at the beginning, as, if the neutralization is effected too slowly, tin is apt to remain unprecipitated. When the action ceases pour a little of the liquor into a beaker containing some hydrogen sulphide water; if the precipitate is white all the tin has been thrown down. Pour away the mother liquor as closely as possible, after allowing any floating particles of metal to settle. The best plan is to pour the liquid from the flask into a beaker, which is then rapidly filled with water from the tap. Press the larger pieces of spongy tin against the side of the beaker with a glass rod, after which they will readily settle. The addition of a drop or two of ammonia, or pouring the liquid to and fro from one beaker to another, will generally insure the settling of the lighter particles. Wash the metal in the beaker back into the flask with 180–200 cc. of hydrochloric acid, to dissolve the tin and zinc, and bring the liquid to a boil as rapidly as possible, with the addition of a granule of zinc. The flask should be provided with a rubber stopper through which passes a short length of glass tubing. If the acid is not brought to a boil rapidly the metal may be completely dissolved while there is still air in the flask, and some stannic chloride be formed. When the solution is complete, and the liquid boiling, titrate at once with ferric chloride. One drop of the ferric chloride solution in excess should produce a plain yellow tinge. The end-point is more marked the hotter and more acid the solution.

Under ordinary circumstances it is better to let the action cease completely when precipitating with the zinc, but in the rare cases when titanium is present (as evidenced by the violet color of the solution during the precipitation), the liquor must be poured off from the tin and zinc while it is still slightly acid, the titanium being then all in solution as sesquichloride. Also, the sponge of tin and zinc should be washed once with water slightly acidulated with hydrochloric acid. Under these circumstances no titanium hydroxide is left with the tin and zinc. Ti_2Cl_4 is oxidized by ferric chloride, though it is unaffected by iodine. In the case of an insoluble silicious residue from solution of the ore after gas reduction, which is apparently poor in tin and appears to be rich in titanium, it would be a convenient plan, after fusing and dissolving in hydrochloric acid, to reduce the solution with iron and titrate with iodine, instead of precipitating with zinc for the ferric chloride assay, as the silicic acid sometimes deposited from hydrochloric acid solutions during zinc precipitation may carry down titanic acid with it or occlude titanous chloride mechanically. If, however, any titanium got through into the hydrochloric acid solution for ferric chloride titration, its presence would at once be shown by the color. On rich material the ferric chloride titration should be employed whenever possible, as the iodine titration, even under the best conditions, has a slight tendency to give low results.

12. *Preparation and Standardization of the Ferric Chloride Solution.*—The ferric chloride solution should be made up 2 liters at a time, from a concentrated stock solution made by dissolving piano wire in hydrochloric acid, peroxidizing with nitric acid and evaporating twice with hydrochloric acid to dryness, and then dissolving in hydrochloric acid. It is standardized against 1 gram of the purest tin, filed wth a fine file. The ferric chloride solution may also be made by dissolving 180 grams of the yellow commercial lump salt, which is roughly $Fe_2Cl_6.12H_2O$, in about 200 cc. of hydrochloric acid and evaporating to dryness. The residue is dissolved in 300 cc. of hydrochloric acid and diluted to 2 liters. The evaporation is to free the solution from nitric acid and arsenic. Use only the purest and strongest hydrochloric acid.

To standardize, weigh 1 gram of pure finely divided tin into an 8-oz. flask, add strong hydrochloric acid until the flask is about three-fourths full, insert a rubber stopper provided with a few inches of small glass tubing, and boil the mixture until solution is complete. Do not boil too rapidly, as this weakens the hydrochloric acid too much before the tin is dissolved. Neither should the liquid come to a boil too slowly, as the tin might then be all dissolved before the air in the flask was thoroughly expelled, and stannic chloride be formed. When solution is complete, titrate the boiling-hot solution at once. Use a fast-running burette with a glass stopcock. Two or three minutes delay in titrating will not show a perceptible effect, but after five minutes, oxidation is apparent and increases rapidly. The end-point of the titration is indicated by a yellow tinge, which is produced by an excess of 1 drop of the ferric chloride solution. If the liquid in the flask turns dark greenish after the titration, the ferric chloride is contaminated with nitric acid. 1 cc. should equal about 2 per cent. of tin, on the basis of one 1 gram taken for assay.

In the assays the main portion of the titration must be done with the above ferric chloride solution, but the finish may be with a solution of half or quarter strength, in order to obtain a more exact ending.

Notes.—Any copper in the tin ore is found in the nitric acid solution, though traces may remain with the SnO_2. In the iron wire stage, the arsenic which escaped extraction with nitric acid and volatilization in the reduction tube, is partly evolved as AsH_3 and partly precipitated with antimony in the metallic form. It comes down as a brown flocculent deposit, which contains 3 or 4 per cent. of its weight in tin. As there is generally only 1 or 2 per cent. at most, and usually under 0.5 per cent., of arsenic in a tin ore, the loss of tin from this cause is quite negligible, but as a check one should save the iron wire and precipitates and filter-papers, and examine them from time to time for tin. It will be found, as in the case of the deposit in the tube, and the dilute hydrochloric acid through which the escaping gas bubbles, that only the merest traces of tin are lost in these operations. Further, the hydrochloric acid solution of the reduced metal may be done in a conical flask with a rubber stopper and tube dipping under water, to assure oneself that there is no appreciable loss by volatilization of $SnCl_2$.

If the ore contains wolfram it is mostly found as WO_3 in the residue from the hydrochloric acid and nitric acid extraction of the reduced metal, from which it may be removed, before fusion, with ammonia. Any tungsten which gets into the main solution comes down as blue oxide with the iron wire precipitate, and any which is fused with sodium carbonate and borax should be removed by reducing the hydrochloric acid extract of the melt with iron wire before precipitating with zinc. In general, all the antimony and some of the lead in the ore will be found in the main hydrochloric acid solution, while some of the lead will be obtained in the nitric acid extract.

After the nitric acid evaporation the residue may be boiled with hydrochloric acid (about 50 cc.), diluted and filtered, though in this case the extract must be tried for tin as a matter of precaution. It will, in general, hold all the copper and most of the arsenic, antimony, lead and iron, though one can never be sure that the residue is free from the oxides of these metals. Occasionally this hydrochloric acid extract will hold a little tin. The residue is reduced in the usual manner.

Except with impure ores the reduction with iron wire may be omitted, as it is always an advantage to save a filtration when possible, and as the small amount of antimony usually present is evolved as stibine during the zinc precipitation. In this case the aliquot part of the solution is at once precipitated with zinc. The boiling hydrochloric acid solution for titration with ferric chloride must, however, be free from black powder of Cu, Sb, etc., and from tungsten blue in suspension or solution. (Lower oxides of tungsten in solution give a brownish pink or light claret colored solution.) During precipitation with zinc the solution should be freely acid to start with, and in the presence of Ti or Mo should be distinctly acid when poured off, as titanium and molybdenum interfere with the ferric chloride titration, titanium sesquichloride reducing ferric chloride.

When copper, or appreciable amounts of antimony, or arsenic, or bismuth are present, reduction from iron must precede zinc precipitation. Instead of doing this with iron wire in a flask use a 400-cc. beaker (tall shape), and one strip of sheet-iron about ½-in. in width, and of sufficient length to leave about 1 inch above the surface of the liquid. This will be easy to wash. The assay must be filtered hot, and the paper thoroughly but rapidly washed with hot water strongly acid with hydrochloric acid. Under the assay conditions the residue is free from tin. After zinc precipitation care must be taken to insure that no floating particles of metal are poured off with the mother liquor, or left in the beaker after redissolving in hydrochloric acid. The mother liquor must always be tested with hydrogen sulphide. The assay for titration is brought to a boil as rapidly as possible, and should begin to boil before the metal is all in solution. The addition of a granule of zinc effects this. The zinc used in precipitation must be good thin sheet-zinc, free from tin.

CHAPTER XXVII.

TITANIUM.

1. **Method for Iron Ores.***—Treat 5 grams of the ore with 30 cc. of strong hydrochloric acid. Heat gently until as complete a decomposition as possible is effected and then evaporate to dryness. Take up the residue in 20 cc. of strong hydrochloric acid, dilute sufficiently and filter off the insoluble residue. The titanium will be found in both the residue and filtrate; they are, therefore, treated separately as follows:

Residue.—Dry and ignite the residue in a platinum crucible so as to burn off all the carbon of the filter. Cool, moisten with water, add 5 to 10 drops of strong sulphuric acid and enough hydrofluoric acid to dissolve the silica, and heat cautiously until fumes of sulphuric acid are given off. Reserve the crucible and contents for treatment described later.

Filtrate.—Heat to boiling in a large beaker. Remove from the heat and add gradually a mixture of 10 cc. of a strong solution of ammonium acid sulphite (see foot-note, p. 130) and 20 cc. of ammonia, stirring constantly. A precipitate forms at first which redissolves. If the precipitate fails to redissolve by vigorous stirring, at any time during the addition of the sulphite, add a few drops of hydrochloric acid to clear the liquid and then continue with

---

* Adapted from Blair. Chem. Anal. of Iron.

the sulphite. When all but 2 or 3 cc. of the sulphite solution has been added replace the beaker over the lamp. The solution should smell quite strongly of sulphur dioxide. Now add ammonia water, drop by drop, until the solution is quite decolorized, and finally a *slight* greenish precipitate is formed which remains even after vigorous stirring. When this occurs add the remaining 2 or 3 cc. of the sulphite solution. This should produce a white precipitate which usually redissolves, leaving a clear and almost decolorized solution. Should, however, any precipitate remain undissolved, add hydrochloric acid, drop by drop, until the solution is clear. It should smell perceptibly of sulphur dioxide. If the reagents are used in exactly the proportions indicated, the reactions will take place as described and the operations will be readily and quickly carried out. If the ammonium acid sulphite solution is weaker than it should be, of course all the ferric chloride may not be reduced and the solution at the end of the operation described above may not be decolorized or smell of sulphur dioxide. In that case add more of the ammonium acid sulphite solution, without ammonia, until the solution smells strongly of sulphur dioxide, and then add ammonia until the slight permanent precipitate appears, and redissolve it in as few drops of hydrochloric acid as possible. The solution being now nearly neutral, the iron in a ferrous condition and an excess of sulphur dioxide being present, add 5 cc. of strong hydrochloric acid to make it decidedly acid and insure complete decomposition of any excess of ammonium acid sulphite which may be present. Boil the solution while passing a stream of carbon dioxide through it until every trace of sulphur dioxide is expelled. Now add a few drops of bromine water and cool the solution by placing the beaker in cold water. To the cold solution add ammonia water from a small beaker very slowly, finally drop by drop, with constant stirring. The green ferrous hydroxide which forms at first dissolves on stirring to a perfectly clear solution, but subsequently, although the green precipitate dissolves, a whitish one remains and the next drop of ammonia increases the whitish precipitate or gives it a reddish tint. Finally the greenish precipitate remains undissolved, even after vigorous stirring, and another drop of ammonia makes the whole precipitate appear green. If, before this occurs, the precipitate does not appear decidedly red in color, dissolve the green precipitate by a drop or two of hydrochloric acid, add 1 or 2 cc. of bromine water, then ammonia as before. Repeat this until the reddish precipitate is obtained, followed by the green coloration as described above. Dissolve this green precipitate in a very few drops of acetic acid (sp. gr. 1.04), when the precipitate remaining will be quite red in color. Now add 1 cc. of acetic acid and dilute the solution with boiling water so that the beaker may be about four-fifths full. Heat to boiling, boil 1 minute, lower the flame and filter as rapidly as possible through a 5½-inch filter. Wash once with hot water. The filtrate should run through clear, but in a few minutes it will appear cloudy owing to the precipitation of ferric hydroxide formed by the action of the air. The points to be observed are the red color of the precipitate and the clearness of the solution when it first runs through.

Dry the filter and precipitate in the funnel. Clean out any precipitate still adhering to the beaker by wiping it with filter-paper and dry this paper. Transfer the thoroughly dried precipitate to a small porcelain mortar, removing it from the filter-paper as completely as possible. Roll up the filter and other bits of paper used and burn them on the lid of the crucible in which the original residue was treated and transfer the ash to the mortar. Grind the precipitate and ash with 3 to 5 grams of sodium carbonate and a little sodium nitrate and transfer it to the crucible containing the residue which was treated with hydrofluoric and sulphuric acids. Clean the mortar and pestle by grinding a little more sodium carbonate and add this to the rest in the crucible. Fuse the whole for half an hour or more, cool, dissolve the fused mass in hot water and filter from the insoluble ferric oxide, etc. This residue contains all the titanium in the form of sodium titanate. (The filtrate contains all the phosphorus in the ore as sodium phosphate.) Dry the residue, transfer it to the crucible in which the last fusion was made, burn the filter and add the ash, and then fuse the whole with 5 grams of dry sodium carbonate. Allow the crucible to cool and then pour strong sulphuric acid into it very gradually. When the effervescence slackens, warm the crucible slightly and continue the gradual additions of sulphuric acid and the careful applications of increased heat until the mass becomes liquid and the ferric oxide is all dissolved. Now heat very carefully until copious fumes of sulphuric anhydride are given off, then allow the crucible to cool and pour the contents, which should be just fluid when cold, into a beaker containing about 250 cc. of cold water. Add to it about 50 cc. of a strong aqueous solution of sulphur dioxide, or 2 or 3 cc. of a strong solution of ammonium acid sulphite. Filter if necessary, nearly neutralize with ammonia, allow to stand until entirely decolorized and then add a filtered solution of 20 grams of sodium acetate in one-sixth the volume of the solution of acetic acid of 1.04 sp. gr. and heat to boiling. The titanic acid will be precipitated almost immediately in a flocculent condition, nearly or quite free from iron. Boil for a few minutes, allow the titantic acid to settle, filter, wash with hot water containing a little acetic acid, dry, ignite, and weigh as TiO_2. Multiply this weight by 0.6005 to obtain that of the titanium.

If the TiO_2, after ignition, appears discolored, owing to the presence of a little ferric oxide, fuse it with a little sodium carbonate, add sulphuric acid to the cold fused mass, dissolve, and repeat the precipitation with sodium acetate in the presence of sulphur dioxide and acetic acid, precisely as before.

2. **Method for Pig Iron, etc.**—Treat 5 grams of drillings in a covered beaker with 80 cc. of nitric acid of 1.2 sp. gr. When violent action has ceased add 10 cc. of strong hydrochloric acid. Remove the cover and evaporate to dryness. Replace the cover and heat until the ferric nitrate is about all decomposed. Cool, add about 30 cc. of hydrochloric acid, heat gently until the iron oxide is dissolved and evaporate to dryness again, best in an air-bath. Cool, dissolve in 30 cc. of dilute hydrochloric acid, and then filter and wash with hot water.

Treat the residue and filtrate separately, precisely as described above for ores (1).

3. **Colorimetric Method for Iron Ores.***—Take 0.1 gram of the finely powdered mineral and mix it in a platinum crucible with 0.2 gram of sodium fluoride, also finely powdered. Add 3 grams of sodium pyrosulphate without mixing. Fuse carefully, holding the burner in the hand. Heat gently till the effervescence ceases and copious fumes of sulphuric acid are evolved. This takes only 2 or 3 minutes. When cold, dissolve the mass in the crucible in 15 to 20 cc. of cold water and filter the solution. The filtrate and washings need not exceed 30 cc. If a residue remains, it can be treated again by the same method, after burning the filter, but the amount of titanium found by a second fusion is usually very small.

To the solution add 1 cc. of hydrogen peroxide and a few cubic centimeters of dilute sulphuric acid and compare the color

---

* Method of W. A. Noyes, Jour. of Anal. and Appl. Chem., V, 39.

(orange-red to yellow) with that of solutions containing known amounts of titanium. Nessler tubes may be used for this purpose, and the solutions are all brought to the same volume, say 30 or 50 cc. For a standard solution, dissolve titanic oxide in hot concentrated sulphuric acid and dilute the solution till 1 cc. contains 1 mg. of TiO_2. In diluting it is best to use dilute sulphuric acid at first, to prevent the precipitation of titanic oxide.*

The colors produced are more or less affected by the presence of iron, and it is therefore advisable to add to the comparison-tubes an amount of ferric sulphate corresponding approximately to that in the solution which is being tested. A solution of ferric ammonium alum answers well for this purpose, and all that is necessary is to match the color of the solution of the mineral before adding hydrogen peroxide to it. If this is done, titanium can be readily determined in the presence of very considerable amounts of iron. Thus, 0.02 mg. of titanic oxide can be detected in 30 cc. of water in the presence of 0.1 gram of ferric oxide in the form of sulphate. This would correspond to 0.02 per cent. for 0.1 gram of mineral.

Determinations based on a comparison of tints are especially valuable for the estimation of small quantities of elements, and for most cases where titanium requires determination the above method is amply accurate. It works excellently with magnetite and other iron ores. There appears to be no appreciable error caused by the volatilization of titanium as fluoride.

4. A qualitative test for titanium can be made in 5 minutes as follows: Mix a little of the powdered ore with sodium fluoride, add sodium pyrosulphate, and fuse as above. Cool by dipping the crucible in cold water. Add 2 or 3 cc. of dilute sulphuric

---

* See 5.

acid and 10 cc. of water. Dissolve by boiling. Divide the solution in two portions, and to one add a few drops of hydrogen peroxide. A comparison with the solution to which no hydrogen peroxide has been added will show at once whether titanium is present or not.

5. **Preparation of Standard Solution of Titanic Oxide.**—I have found difficulty in preparing a standard solution of titanium by dissolving titanic oxide in hot concentrated sulphuric acid. The following method is satisfactory:

Ignite pure titanic acid until all water of hydration is expelled. Weigh 0.5 gram into a 6-oz. flask, add 10 cc. of strong sulphuric acid and 5 grams of potassium sulphate. Heat over a free flame until the titanic acid is dissolved and the solution is clear. Cool, dilute with 125 cc. of 1:4 sulphuric acid, transfer to a 500-cc. measuring flask, and make up to the mark with cold water. When perfectly cool, again adjust to the mark and mix thoroughly. 1 cc. will contain 1 mg. of TiO_2. A more accurate method of preparing this standard solution may be found in Bulletin No. 305 of the U. S. Geological Survey, p. 111.

CHAPTER XXVIII.

TUNGSTEN.

(See Appendix.)

THE following method for the determination of tungsten in wolframite and oxidized ores is modified from one devised by O. P. Fritchle.

1. Hydrofluoric Acid Method.—Treat 0.5 gram of the very finely powdered ore in a small platinum dish with equal parts of strong hydrochloric and hydrofluoric acids. Digest on a water-bath until solution is complete, adding more of each acid from time to time if necessary. It may require from one to several hours to effect complete decomposition of the ore. Usually a perfect solution may be obtained. Any tin oxide present will be unaffected. Finally, evaporate to about 15 cc. with an excess of hydrochloric acid. A yellow precipitate of H_2WO_4 may separate during the final evaporation, owing to the expulsion of the hydrofluoric acid that holds it in solution. This will do no harm provided it can all be removed from the dish. Transfer the solution and any precipitate to a 6-oz. flask, add 20 cc. of strong hydrochloric acid and 8 cc. of strong nitric acid. Boil down to about 10 cc. This will expel any remaining hydrofluoric acid and precipitate the tungsten as tungstic acid, H_2WO_4. Dilute with 50 cc. of hot water and allow to simmer at a gentle heat for about half an hour, or until the tungstic acid has settled clear. Filter, wash well with hot water slightly acidulated with hydrochloric acid, and then dissolve the tungstic acid on the filter by pouring warm dilute ammonia over it, using as little as possible, and washing the filter with the same solution. Receive the filtrate in a weighed platinum dish. Evaporate the solution on a water-bath to dryness, and then ignite the residue at a red heat, cool and weigh as WO_3. The cold residue should be of a bright canary-yellow color. Multiply the weight of the WO_3 by 0.793 to obtain the weight of the tungsten, from which the percentage in the ore may be calculated.

2. **Fusion Method for Wolframite.**—Fuse 0.5 gram of the very finely ground ore with 2–3 grams of sodium potassium carbonate in a platinum crucible for from one-half to three-quarters of an hour. Dissolve the fused mass in boiling water. The tungsten goes into solution as sodium or potassium tungstate, together with alkali silicate and also stannate, if tin be present. The residue contains the iron, manganese, calcium, and magnesium. Filter and wash with hot water. Rinse the residue into a beaker and warm with dilute hydrochloric acid. If gritty particles remain undissolved filter them off through the filter last used and wash with hot water. Dry and ignite the residue and again fuse it with the mixed carbonates. Dissolve the fused mass as before, filter and unite the filtrate with the first filtrate.

3. Having thus obtained an aqueous solution of alkali tungstate, add to it an excess of nitric acid and evaporate to dryness on a water-bath. Again add a little nitric acid and evaporate to dryness a second and third time. Finally heat the residue in a drying-oven at 120° C. for some time, then moisten with strong nitric acid and allow to stand for 15 or 20 minutes. Now add a hot 5 per cent. solution of ammonium nitrate and filter the mixture, washing well with ammonium nitrate solution slightly acidified with nitric acid to remove all the sodium and potassium salts. Finally, wash once or twice more with a hot,

very dilute ammonium nitrate solution and then dry the filter and contents and transfer the latter as completely as possible to a weighed platinum crucible. Moisten the paper with a strong solution of ammonium nitrate, dry it and incinerate over the crucible in a coil of platinum wire. Ignite the whole, now, with free access of air. If the tungstic acid is not pure yellow when cool, moisten with a few drops of nitric acid and repeat the ignition.

4. The ignited tungstic acid may contain silica and stannic oxide. The former may be removed by warming with a few cubic centimeters of hydrofluoric acid, evaporating to dryness and igniting. The residue consists of pure tungstic acid, or tungstic acid and stannic oxide. The amount of the latter is usually so small as to be negligible. If desired, however, the tin may be volatilized as stannic chloride by ignition with ammonium chloride. The stannic chloride is decomposed by the moisture of the air and stannic oxide may be deposited on the outside of the crucible. To prevent this, place the crucible in a larger one and keep the outer crucible covered until the ammonium chloride is completely expelled. Now heat the inner crucible with free access of air until its contents become of a pure yellow color. Cool and weigh. Repeat the ignition with six or eight times as much ammonium chloride as there is precipitate until the weight of the residual WO_3 becomes constant. The tungstic acid becomes dark on igniting in the absence of air and only assumes its true color and weight on igniting with free access of air.

The weight of the WO_3 multiplied by 0.793 gives that of the tungsten.

5. Another method of precipitating the tungstic acid from the solution of alkali tungstate is that of Berzelius, as follows: Neutralize the greater part of the alkali with nitric acid, being very careful, however, to still leave the solution slightly alkaline. The amount of nitric acid to use is best ascertained by a blank

test on the same amount of alkali carbonates as was taken for the fusion. Now add a solution of mercurous nitrate until it produces no further precipitation. If, on slowly adding the mercurous nitrate, the precipitate seems to be getting unduly large, indicating too great alkalinity, add nitric acid drop by drop until an added drop of mercurous nitrate produces no cloud. Heat to boiling, allow the precipitate to settle, then filter and wash with water containing mercurous nitrate. Dry the filter and contents and then ignite in a platinum crucible under a hood. Weigh as WO_3.

The tungstic acid thus obtained almost always contains silica and possibly stannic oxide. It may be purified as described above (4).

6. The Determination of Tungstic Acid in Low-Grade Ores.*— Weigh 5 grams or more of the sample. Digest in a 4-inch porcelain dish with 20 cc. of a 25 per cent. solution of sodium hydroxide (free from chloride) for 30 to 45 minutes on a water-bath. Now dilute the mixture somewhat, add a little sodium peroxide to oxidize any decomposition products of sulphides, transfer to a 250-cc. measuring-flask and make up to the mark. After mixing, filter through a dry filter, reject a little of the first filtrate that runs through, and collect 200 cc. of the remainder in a measuring-flask. Transfer this portion to a beaker, acidify with nitric acid and then make alkaline with ammonia. Heat to boiling, filter and wash. Acidify the filtrate slightly with dilute nitric acid, add mercurous nitrate solution (prepared as below) in excess, and then a few drops of dilute ammonia. On warming and stirring, the precipitate settles readily. Filter, wash the precipitate with weak mercurous nitrate solution, and then ignite the precipitate and paper together in a weighed porcelain crucible.

* H. W. Hutchin and F. J. Tonks, Inst. Min. and Met. Bull., No. 56.

A platinum crucible may be used if the ore is free from arsenic. Weigh as WO_3.

For assays of ores and tailings, the sample may be reduced to a sufficient degree of fineness in a wedgewood mortar, but for concentrates an agate mortar is necessary; fine powdering is essential. For decomposing charges containing not more than 0.4 gram of tungstic acid, 20 cc. of a 25 per cent. solution of sodium hydroxide will suffice. The decomposition is rapid; 0.4 gram of wolfram concentrates being decomposed in 15 minutes to the extent of 98 per cent. of the tungsten content. As a rule, however, it is best to take from 30 to 45 minutes for the decomposition.

The mercurous nitrate solution may be conveniently prepared from mercury. Digest from 2 to 3 ozs. in a large beaker or flask for 1½ hours with 25 cc. of nitric acid (sp. gr. 1.4) and 75 cc. of water, on a hot plate which will keep the liquid nearly at the boiling-point. Allow to stand, hot, over night. Dilute to about 400 cc., which will give a saturated solution with a minimum of free acid.

The weighed WO_3 may be tested for silica, if desired, by treatment in platinum with hydrofluoric acid. The loss is usually very slight.

In the presence of scheelite the method is not applicable, since this mineral is only partially attacked, under the conditions of the assay.

7. **Method by Fusion with Alkalies.**—Treat 5 grams of the finely powdered ore with hydrofluoric acid in a platinum dish, finally evaporating to dryness. Wash the dried residue into a beaker with water, to a volume of about 40 cc. Add 5 cc. of nitric acid and warm gently to remove the bulk of any mispickel present and then dilute considerably and allow to stand over night. Filter the mixture and wash, dry, and ignite the residue.

Again powder the residue in an agate mortar, transfer it to a nickel dish and fuse it with a mixture of sodium peroxide and hydroxide. Extract the melt, when cold, with water and dilute to 250 cc. in a measuring-flask. Mix well and filter through a dry filter, rejecting the first portion. Measure 200 cc. of the filtrate (representing 4 grams of ore) to be used for the actual assay. Transfer to a beaker, acidify with nitric acid, then make plainly alkaline with ammonia, boil, filter, and wash. Add nitric acid to the filtrate until it is neutral or faintly acid, then add mercurous nitrate solution (described in 6) in excess, and finally a little freshly precipitated mercuric oxide, or a few drops of dilute ammonia. Warm and stir the mixture until it settles well and then filter, washing the precipitate with weak mercurous nitrate solution. Ignite paper and precipitate in a platinum crucible and weigh the residue as WO_3. The WO_3 should be tested for silica by treatment with hydrofluoric acid.

CHAPTER XXIX.

URANIUM AND VANADIUM.

(See Appendix.)

1. **Method for Uranium.**—Treat 1 gm. of the ore (or more if very low grade) in an 8-oz. flask with 10 cc. of strong hydrochloric acid and 5 cc. of strong nitric acid. Allow to simmer gently over a low heat until solution is as complete as possible, and then boil to dryness. This may be done by manipulating the flask in a holder over a free flame. Add 3 cc. of hydrochloric acid and 5 cc. of water to the residue, warm for a short time, occasionally agitating, then dilute with 25 cc. of hot water and filter, washing with warm water. Receive the filtrate in a small beaker.

2. Pass hydrogen sulphide into the liquid to separate copper, lead and other metals of this group, filter, wash with hydrogen sulphide water and boil the filtrate to expel the hydrogen sulphide. Concentrate to 100 cc. if necessary, oxidize with hydrogen peroxide (usually 10 cc.), and then neutralize with dry sodium carbonate, adding 2 or 3 grams in excess. Boil the liquid for about 15 minutes, or until the yellowish uranium precipitate dissolves, leaving a brown precipitate which is largely iron. Filter and wash the iron precipitate once or twice with hot water, and reserve the filtrate. Dissolve the iron precipitate in the least possible amount of nitric acid (1 : 1), dilute somewhat if necessary, add 10 cc. of hydrogen peroxide and repeat the precipitation with sodium carbonate precisely as before. Filter into the beaker containing the first filtrate. Wash well with hot water.

3. Evaporate the united filtrates from the iron precipitation to a volume of about 200 cc., add 10 cc. of strong nitric acid and boil until all CO_2 is expelled. Neutralize the free acid with ammonia (until a slight pernament precipitate appears), then add 4 cc. of nitric acid for each 100 cc. of liquid. Now add 10 cc. of a 20 per cent. lead acetate solution, and enough of a strong solution of ammonium acetate to neutralize the nitric acid present and substitute acetic acid for it. The object is to precipitate the vanadium as lead vanadate in an acetic acid solution. The ammonium acetate solution may be made by mixing 80 cc. of strong ammonia, 100 cc. of water, and 70 cc. of acetic acid 99 per cent. pure.

4. Heat the liquid containing the lead vanadate precipitate on the steam-bath for one hour or more, filter on a tight filter and wash with warm water. Dissolve the precipitate in the least possible quantity of hot dilute nitric acid, neutralize as before, add 3 cc. of nitric acid in excess, then 2 cc. of lead acetate solution and repeat the precipitation of lead vanadate by adding ammonium acetate in excess. Filter and add the filtrate to the one from the first precipitation of lead vanadate. Evaporate the united filtrates from the lead vanadate to about 400 c.c Add 10 cc. of strong sulphuric acid to separate the bulk of the lead (derived from the excess of lead acetate) as $PbSO_4$, and filter, washing the precipitate with cold water. Neutralize the filtrate from the $PbSO_4$ with ammonia and add freshly prepared $(NH_4)HS$ until the solution is yellow and the uranium and what little lead is present are precipitated as sulphides. Warm up mixture gently until the sulphides settle well. Filter and wash slightly with warm water.

5. Dissolve the precipitate in a No. 2 beaker with hot dilute (1 : 2) nitric acid, add 5 cc. of strong sulphuric acid and evaporate until the latter is fuming. Cool, take up with water, boil, let

the small precipitate of $PbSO_4$ settle until the liquid is cold, and then filter it off, washing with very dilute sulphuric acid (1:20).

6. *Separation of Alumina:* Nearly neutralizes the filtrate with ammonia. Have the solution cold (not warmer than 30° C.), and add powdered ammonium carbonate in about 2 grams excess. This will precipitate the aluminum and hold the uranium in solution. Let the precipitate settle, filter, and wash it with warm water. If the precipitate is bulky, or is at all yellow, dissolve it in a little dilute sulphuric acid and reprecipitate with ammonium carbonate as before.

7. Acidify the filtrate, or combined filtrates, from the alumina with sulphuric acid and boil to thoroughly expel CO_2. Make the liquid slightly alkaline with ammonia while it is hot, and heat gently until the ammonium uranate collects and settles. Filter and wash with a very dilute solution (2 per cent.) of ammonium nitrate. Do not allow the precipitate to run dry on the filter after the first washing. Dry the precipitate, ignite it in a porcelain crucible and weigh as U_3O_8. Dissolve the ignited residue in a little nitric acid and test it with H_2O_2 for vanadium. Only a faint brownish tint should appear, at most. Rinse the solution into a small beaker and test for alumina with ammonium carbonate. Should an appreciable amount be found it may be filtered off, ignited, weighed and deducted.

8. **Volumetric Method for Uranium.**—Treat 1 gram of the ore (or more if very low grade) in an 8-oz. flask with 10 cc. of strong hydrochloric acid and 5 cc. of strong nitric acid. Allow to simmer gently over a low heat until solution is as complete as possible, and then boil to dryness. This may be done by manipulating the flask in a holder over a free flame. Add 3 cc. of hydrochloric acid and 5 cc. of water to the residue and warm

gently until solution is as complete as before. Now add 35 cc. water and pass in hydrogen sulphide to precipitate the members of that group. Filter, washing with hydrogen sulphide water at least 7 times, receiving the filtrate in a fairly large beaker. Placing a boiling-rod (p. 8) in the liquid, cover the beaker and boil off the hydrogen sulphide. Remove from the heat and add 10 cc. of H_2O_2. Now add dry sodium carbonate in small portions until the free acid is neutralized and then about 2 grams in excess. Boil the mixture until all CO_2 is expelled and the precipitate, on standing, settles well. At this point add about 2 cc. more H_2O_2. Filter, washing once or twice with hot water, and reserve the filtrate, which may be allowed to boil and concentrate while waiting for the next step. Rinse the bulk of the precipitate from the filter back into the beaker, then place the latter under the funnel and dissolve what precipitate remains on the filter with dilute (1 : 1) nitric acid, using as little as possible. Now wash the filter well with hot water, allowing all to run into the beaker containing the precipitate. Warm the mixture until the precipitate dissolves, adding more acid if necessary, then add H_2O_2 and precipitate with sodium carbonate precisely as before. Filter and wash the precipitate well with hot water. Add the filtrate to the first one, which has been concentrating, and boil the whole to a bulk of about 100 cc. Now remove from the heat (simply to prevent frothing over) and cautiously add 10 cc. of strong nitric acid, and then boil again until all the CO_2 is expelled. The solution must be clear and boiling hot for the next step.

9. *Separation of Vanadium.*—Place in a large beaker about 15 grams of ammonium acetate, 5 grams of microcosmic salt, 50 to 75 cc. of water, 5 cc. of glacial acetic acid. Place a boiling-rod in the mixture and heat to rapid boiling. Have arranged over the beaker a funnel with the lower end drawn out so as

to deliver only a fine stream. Pour the hot uranium solution through this into the boiling phosphate solution. Allow to boil a few minutes after all has run through. Now remove the beaker from the heat, cover it and allow to stand until the precipitate has settled well and then filter. Time will be saved by pouring as much as possible of the clear liquid through the filter before disturbing the precipitate. Wash the precipitate only once, using hot water. Now rinse it back into the beaker, place the latter under the funnel and dissolve what remains on the filter with a little hot, dilute nitric acid. 6 cc. of 1 : 1 acid are usually sufficient. Wash the filter well with hot water, receiving all in the beaker containing the precipitate. Warm the mixture until the precipitate dissolves and dilute to about 75 cc, if necessary. Heat to boiling and repeat the phosphate precipitation precisely as before. The two operations will remove practically all the vanadium and leave the uranium and aluminum as phosphates. Filter as before and wash four or five times with hot water containing a little ammonium sulphate. This is tedious if there is much aluminum present. Time is saved by stirring the precipitate in the apex of the filter with the jet each time. Use the same filter as before.

10. Make a mixture, in a small beaker, of 5 cc. of strong sulphuric acid and 15 cc. of water, and heat nearly to boiling. Place an 8-oz. flask under the funnel and pour the hot acid mixture slowly over the precipitate on the filter. With care, it may thus be all dissolved, although the hot acid may be poured through again if necessary. Wash the filter well with hot water, but do not unduly increase the volume of liquid in the flask. This should not be allowed to much exceed 50 to 75 cc. Heat nearly to boiling and run in permanganate to a permanent deep pink color, to oxidize any organic matter

present. Now add 10 grams of 20-mesh C. P. zinc and allow to stand until the action has become very slow, then add 25 cc. more water and 10 cc. more sulphuric acid. When the zinc has nearly all dissolved, filter the solution through a fresh filter, washing the latter and residue well with cold water. Receive the filtrate in a 16-oz. flask. Dilute with cold water, if necessary, until the flask is about half full.

11. Cool the solution to room temperature and titrate with the ordinary permanganate solution used for iron (p. 119) until a single drop produces a pink color that lasts several seconds. The end-point is very sharp, although the color may soon fade, owing, possibly to traces of vanadium. Note the burette-reading and deduct the correction required for the iron in the 10 grams of zinc used, this having been determined by a blank test. As a rule, no further correction is required, as any trace of vanadium present appears to be acted on so slowly in the cold as not to seriously interfere with obtaining a sharp end-point. As a precaution, however, add 1 gram of sodium sulphite and boil the solution in the flask until all smell of SO_2 has gone, and then 7 or 8 minutes more. Add cold water to restore the original volume, cool to room temperature and titrate as before. Run the above zinc blank in the same way and find the correction required for the combined iron and impurities in the sulphite. After deducting this from the last reading, the remainder, if any, may be credited to vanadium and deducted as a further correction from the uranium reading.

12. 55.85 Fe $=$ 140.583 U_3O_8, or 1 Fe $=$ 2.517 U_3O_8. If the percentage value of 1 cc. of the permanganate for iron, based on 0.5 gram of ore taken, is given, this, multiplied by 1.2585, will give the percentage value of 1 cc. for U_3O_8, when 1 gram of uranium ore is taken.

13. **Method for Vanadium.***—Treat 1 gram of the finely divided ore in a platinum dish with about 3 cc. of strong sulphuric acid and 20 cc. of hydrofluoric acid. Evaporate to fumes of sulphuric acid and then expel the latter by heating over a naked flame. Now add about 2 grams of sodium carbonate and fuse. Extract the melt with hot water and filter, washing with hot water. Acidify the filtrate with sulphuric acid, heat nearly to boiling and pass in hydrogen sulphide gas. Arsenic and molybdenum are precipitated, if present, and V_2O_5 is reduced to V_2O_4. Filter, washing with hydrogen sulphide water. Boil the filtrate until the hydrogen sulphide is completely expelled and then titrate the hot solution with a standard solution of potassium permanganate to the usual pink tinge (XV, 8). Then again reduce by passing in sulphur dioxide gas, boil off the excess and repeat the titration. The latter result is apt to be a little lower than the former, and is to be taken as the correct one. The iron factor of the permanganate multiplied by 0.916 will give the vanadium factor; care being taken to use the absolute iron factor and not the percentage factor based on 0.5 gram of ore being taken for analysis. Uranium does not interfere with this method, which I regard as the best of the many technical methods proposed.

14. The direct use of a solution of sulphur dioxide, or an alkali sulphite for reducing the vanadium is inadmissible unless these have been freshly prepared, or a blank is run, since they are liable to contain other oxidizable substances than SO_2 or a sulphite. The SO_2 is best obtained as wanted by heating a flask containing a solution of SO_2, or a sulphite to which sulphuric acid has been added.

15. In case the volume of permanganate used is so small as to make doubtful the presence of vanadium, it is necessary to apply

---

* Adapted from Hillebrand and Ransome, Am. Jour. Sci., x, also Hillebrand, Bulletin 176, U. S. Geological Survey.

a qualitative test, which is best made as follows: The solution is evaporated and heated to expel the excess of sulphuric acid, the residue is taken up with 2 or 3 cc. of water and a few drops of dilute nitric acid, and a couple of drops of H_2O_2 are added. A characteristic brownish tint indicates vanadium. Unless the greater part of the free sulphuric acid has been removed the appearance of the color is sometimes not immediate and pronounced, hence the above precaution. It is also necessary that the nitric acid shall be in considerable excess, since in neutral or only faintly acid solution the color does not appear strongly.

For determining very small amounts of vanadium in rock analysis, the titrated liquid should be from 25 to 100 cc. in bulk and the permanganate solution very dilute, 1 cc. = about 0.001 gram V_2O_5. The temperature for titration should be from 70°–80° C.

16. Volumetric Method for Carnotite.*—This mineral contains uranium and vanadium as potassium uranyl vanadate.

Dissolve a sample of the ore that does not contain more than 0.25 gram of U_3O_8 in sulphuric acid (1:5) and evaporate to fumes of the acid. Cool, dilute, add an excess of sodium carbonate and boil until the precipitate settles well. Filter and wash with hot water. Dissolve the precipitate in the smallest possible amount of sulphuric acid (1:5), dilute, add an excess of sodium carbonate, boil, filter and wash. Acidify the combined filtrates and wash-waters with sulphuric acid. Add ammonium phosphate (0.5 gram is usually sufficient), heat to boiling and make alkaline with ammonia; boil for a few minutes, filter and wash with hot water containing a little ammonium sulphate, which prevents the finely divided particles of the precipitate from passing through the filter.

17. The filtrate now contains the vanadium and the precipitate

* A. N. Finn, Jour. Am. Chem. Soc., XXVIII, 1443.

the uranium. Acidify the filtrate with sulphuric acid, pass sulphur dioxide into it until it becomes blue, boil to expel the excess of sulphur dioxide and titrate while hot with standard potassium permanganate solution. The iron factor of the permanganate multiplied by 1.631 gives the V_2O_5 factor, or, by 0.9159, the vanadium factor.

18. Dissolve the ammonium uranyl phosphate in dilute sulphuric acid, add some granulated zinc and let the action continue vigorously for at least thirty minutes. Remove the undissolved zinc by filtering through asbestos, using a suction pump. (These are Finn's directions; I would suggest a plug of absorbent-cotton placed in the neck of a funnel, without a suction pump, instead of the asbestos.) Titrate the filtrate at about 60° C. with permanganate (twentieth-normal).*

The iron factor of the permanganate solution multiplied by 2.5167 gives the U_3O_8 factor, or, by 2.133, the uranium factor.

19. Method for Vanadium.† — *For Products Low in Silica:* Weigh out from 0.250 to 1 gram of the finely powdered material and mix in a nickel crucible of about 25-cc. capacity with 3 grams of sodium peroxide; cover with about 1 gram more of the peroxide and heat over a Bunsen burner for three or four minutes. It is not necessary to heat to fusion; a dull red being sufficient. When the darkening which gradually takes place on the surface is complete, the operation may be considered finished.

Half fill a 500-cc. beaker with cold water and transfer the cold crucible containing the melt to it. Care should be taken at this point, as the reaction between the peroxide and water is violent. It is well to float the crucible in the beaker, cover the latter with a large watch glass, and add water to the contents

* I would advise making a blank test and deducting the permanganate required.

† H. F. Watts, Western Chem. and Met., V, 408.

of the crucible, a few drops at a time, from a wash-bottle. When solution is complete, remove and wash the cru cible, heat the solution to boiling and pass a current of carbon dioxide through it until saturated. Remove from the heat, allow to settle and filter. Wash 3 times by decantation, and 3 times on the filter, using hot water containing a little sodium carbonate. A platinum cone and suction may be employed if desired.

Make the filtrate just acid with nitric acid and boil the solution a few moments. Remove the beaker from the heat and add to the hot liquid 10 cc. of a saturated solution of lead acetate. If the excess of nitric acid is not too great, a precipitate of lead vanadate will form at once. However, it is necessary to also add a few grams of sodium acetate to insure the complete neutralization of the nitric acid and the absence of any free acid except acetic. The lead vanadate is insoluble in dilute acetic acid, but dissolves readily in nitric acid. Digest the mixture at a gentle heat, or on the water-bath, for a short time, until the precipitate settles well, and then filter and wash with hot water containing a little acetic acid. Rinse the bulk of the precipitate (unless very small in amount) into an 8-oz. flask, place the latter under the funnel and pour through the filter sufficient dilute nitric acid to dissolve any remaining precipitate, and also to dissolve that in the flask, but avoid using more acid than required, or otherwise unnecessarily increasing the amount of liquid to be subsequently evaporated. Add to the solution 10 cc. of strong sulphuric acid and boil to fumes. Cool, dilute sufficiently with cold water and filter off the lead sulphate, washing with cold water. Receive the filtrate in an 8-oz. flask. Boil the filtrate and reduce the vanadium (from V_2O_5 to V_2O_4) by adding 1 gram of sodium sulphite in successive small portions. Boil off the excess of sulphur dioxide, testing for its complete removal as follows: Fit a stopper and delivery tube to the flask and consider the operation ended when the escaping steam no longer decolorizes a dilute solution of potassium permanganate, held in a small beaker. Now titrate the hot solution with standard potassium permanganate to the usual pink tinge, employing the same permanganate as is used for iron titrations. The iron value of the permanganate multiplied by 1.632 will give the V_2O_5 value, or by 0.916 the vanadium value.

If the material contains arsenic it will be necessary to remove it. A convenient point at which to do this is after the reduction with sodium sulphite and expulsion of the excess of sulphur dioxide. Pass hydrogen sulphide, filter, washing with hydrogen sulphide water, boil off the excess of hydrogen sulphide and titrate as before.

The burette reading should be corrected as follows: To 150 cc. of water in a flask, add 10 cc. of strong sulphuric acid and 1 gram of sodium sulphite. Boil off the sulphur dioxide and titrate. The amount of permanganate required to color the solution should be deducted from that used in the assay. The correction usually amounts to 0.2 cc.

Method for Ores: Treat 1 gram of the finely ground ore (or less if high grade) with 10 cc. of aqua regia and evaporate to dryness. Add 10 cc. of nitric acid (1.20 sp. gr.), and digest on the hot-plate for a few minutes. Filter off the silica, etc. Nearly neutralize the filtrate with sodium hydroxide and then pour it with constant stirring into a hot solution of sodium hydroxide contained in a large beaker. A stick of sodium hydroxide dissolved in 200 cc. of water is about the right strength. Heat the solution to boiling and pass a current of carbon dioxide to saturation. Continue from this point as described above.

Method for Products Containing much Silica, and not Readily Decomposed by Acids: The sodium peroxide method may be used, but in this case it will be necessary to remove the silica.

After the filtrate from the iron has been made slightly acid with nitric acid, again make alkaline with ammonia, and add 3 or 4 grams of ammonium carbonate. Place the beaker on a warm plate and allow to stand for an hour with occasional stirring. Filter off the voluminous precipitate of silica, using a 12-cm. filter supported on a platinum cone, and employing suction. This precipitate, though somewhat inclined to be sticky, is not difficult to filter if suction is used. Wash thoroughly with hot water containing a little ammonium carbonate, at least 10 or 12 times, churning up the precipitate well with the stream from the wash-bottle. The filtrate is now freed from the greater part of the silica. Again make it slighly acid with nitric acid and precipitate with lead acetate as before.

When using the acid method of decomposition, it is not always safe to assume that the vanadium is all extracted by treatment of the ore, crushed to 80-mesh or coarser, with dilute nitric acid, as is sometimes recommended. Some vanadiferous sandstones yield to this treatment, but not all. To make sure of complete extraction, the ore should be finely ground in an agate mortar and decomposed first with aqua regia.

The direct reduction and titration of the vanadium in the presence of the accumulated salts of the analysis is inaccurate, the end-point being uncertain, and successive reductions generally failing to check. It is therefore always necessary to separate the vanadium from these impurities with lead acetate, the titration of pure vanadium solutions being exact. If rapid approximate results are wanted, the filtrate from the iron residue may be acidified with sulphuric acid, and the solution reduced and titrated.

The method as described has the advantage over the ammonium carbonate method in that the separation of the iron and vanadium is performed, in nearly every case, in a single operation, and the necessity for redissolving and reprecipitating the iron residue is obviated. A clean separation can be made on iron vanadate, using 300 mg. of substance.

The carbon dioxide is used to prevent bumping (it also neutralizes the caustic alkali, which would otherwise destroy the filter.—A. H. L.), although in some cases its use apparently gives cleaner separations, due possibly to the tendency of the alkaline carbonates formed to hold vanadium in solution.

20. **Note by the Author.**—I would suggest the following procedure for removing the troublesome silica. It appears to be generally applicable:

Treat the ore as described in *Method for Ores* above until the silica has been filtered off. Reserve the filtrate. Ignite the insoluble residue in a platinum dish. After cooling, treat with sufficient strong hydrofluoric acid and evaporate on the water-bath to dryness. Take up in a little dilute nitric acid and again evaporate to dryness, and then once more dissolve in a little dilute nitric acid. Filter, if necessary, adding the filtrate to the one reserved. Ignite the residue in platinum, and then transfer it to an iron or nickel crucible and fuse with a little sodium peroxide. Dissolve the melt in water, add the reserved filtrate, and then sufficient nitric acid or sodium hydroxide solution, as required, to finally leave the solution slightly acid. Proceed with this solution (pouring it into sodium hydroxide solution, etc.) as described in the *Method for Ores* above.

CHAPTER XXX.

ZINC.

1. The following method, developed by the writer, for the technical estimation of zinc in ores, etc., is of general but not of universal application. Ordinarily, however, almost the only source of trouble likely to be encountered is the preliminary decomposition of the ore. The method as described is applicable to the usual run of mixed sulphide ores and many oxidized ores. When the decomposition fails or is doubtful, the operator must note the fact and apply the proper remedy. Some observations on this and related matters are given below.

2. **Author's Method.**—Prepare a solution of potassium ferrocyanide containing 21.55 grams of the crystallized salt to the liter. Standardize this solution as follows: Weigh carefully about 0.2 gram of pure zinc into an 8-oz. flask and add 10 cc. of strong hydrochloric acid (sp. gr. 1.20). When the zinc has dissolved, dilute with about 25 cc. of water, add a few drops of litmus solution as an indicator, and make slightly alkaline with ammonia. Again acidify slightly with hydrochloric acid and then add 3 cc. of the strong acid in excess. Dilute now to about 250 cc. with hot water and heat nearly to boiling. Titrate with the ferrocyanide solution as follows: Pour about two-thirds of the zinc solution from the flask into a 400-cc. beaker. Run the ferrocyanide solution from the burette into this portion, a few cubic centimeters at a time, until a drop, when removed and tested on a porcelain plate with a drop of a 15 per cent. solution of uranium nitrate, shows a brown tinge. Now add more of the

zinc solution from the flask and continue the titration more cautiously until the end-point is again passed. Proceed thus, adding more of the reserved portion in such amounts as may appear advisable, passing the end-point each time with greater caution. Finally, add the last of the reserved portion, and then, to save rinsing out the flask, pour a large part of the solution back into it again and then empty it once more. From this point finish the titration very carefully, ordinarily by testing after each addition of two drops. Instead of using a single drop of the zinc solution for the test, the reaction is sharper if a quantity equivalent to several drops be taken. If this is done only near the end of the titration the amount of zinc lost thereby will be insignificant. A convenient way of making the test is to use a medicine-dropper and place a single drop of the uranium solution in each depression of the test-plate at the outset. By using a glass tube instead of a rod for a stirrer, any desired quantity of the solution can be quickly removed for a test. When the final brown tinge is obtained, note the reading of the burette, and then wait a minute or two and observe if one or more of the preceding tests do not also develop a color. The end-point is always passed by a test or two and the burette reading must be corrected accordingly. A further correction must also be made for the amount of ferrocyanide required to produce a color under the same conditions when no zinc is present. This is ordinarily 1 drop. 1 cc. of the standard solution will equal about 0.005 gram of zinc, or in the case of ores, about 1 per cent. when 0.5 gram is taken for assay.

3. **Regular Method for Ores.**—A modification of the method of decomposition described below is sometimes necessary. Some ores require a more or less prolonged treatment with hydrochloric acid, to insure complete decomposition of oxides, before adding nitric acid. Ores that gelatinize should be mixed with a little

water before adding any acid, and then the acid added while agitating the flask, to prevent setting or caking. This is to be followed by occasional agitation during decomposition. Other cases will be referred to later.

Weigh 0.5 gram of the ore into an 8-oz. Erlenmeyer flask. Add 5 cc. of strong hydrochloric acid and 10 cc. of strong nitric acid. Boil gently until only a few cubic centimeters of liquid remain, but not to dryness in any case. Remove from the heat and add 12 cc. of strong nitric acid and 5 grams (usually measured) of potassium chlorate. (Mallinckrodt's "Pure Granular" is convenient for measuring.) Replace the flask over a gentle heat and run just to dryness. If the mixture shows a tendency to "bump," use a small watch-glass cover raised slightly on one side with a bit of bent glass tubing, to allow escape of vapors, removing it again when the danger is passed. The raising is unnecessary if the flask has a lip. Hard dryness is unnecessary except when gelatinous silica is present, and even in this case there should be no overheating or fusing of the salts. With other ores it is sufficient to run just to dryness and remove at once from the heat. As soon as sufficiently cool add 35 cc. of a prepared ammoniacal solution and heat to boiling. This solution is made by dissolving 200 grams of commercial ammonium chloride in a mixture of 500 cc. of strong ammonia (sp. gr. 0.90) and 750 cc. of water. Boil the contents of the flask very gently for a minute or two, or until disintegration of the residue is complete, and then add saturated bromine water and continue the boiling for a short time. The amount of bromine water to add depends upon the quantity of manganese apparently present, as shown by the deep brown color of the residue, All the manganese is precipitated by the mixture of nitric acid and potassium chlorate, but on evaporating to dryness, some of the manganese is again reduced to the protoxide form and becomes more or less soluble.

With ores showing little or no manganese 10 cc. of bromine water are sufficient, and 25 cc. will usually suffice in any case. Now filter the hot solution through an 11-cm. filter into a 400-cc. beaker. I have found that in most cases the liquid runs through much more rapidly if a small wad of absorbent cotton is placed in the apex of the filter and wetted. Wash out the flask with hot water. A slight residue adhering to the flask may be disregarded; a larger residue that cannot be removed with a rubber-tipped rod may be treated as follows: Add about 2 cc. of strong hydrochloric acid to dissolve it, and then, without diluting the solution, add 5 cc., or an excess, of strong ammonia and rinse into the filter. Any traces of manganese not precipitated by the ammonia are usually negligible. Wash the residue on the filter ten times with a nearly boiling ammonium chloride mixture, which may be contained in the wash-bottle described on page 9. To make the ammonium chloride mixture, dissolve 100 grams of commercial ammonium chloride in water, add 50 cc. of strong ammonia and dilute to 1 liter. Add a little litmus solution to the filtrate as an indicator and then cautiously neutralize with strong hydrochloric acid, finally adding 3 cc. in excess. When there is sufficient copper present to act as an indicator the litmus may be omitted. Dilute the liquid, if necessary, to about 200 cc. with hot water, heat nearly to boiling, and add 50 cc. of saturated hydrogen sulphide water. The mixture is now ready for titration. Pour about one-third of the liquid into a beaker as a reserve and conduct the titration as described for the standardization of the ferrocyanide, adding portions of the reserve as the end-point is successively passed. In correcting for the final reading of the burette, it is usually sufficient to deduct for as many tests as show a brown tinge and one drop additional. This is shown by running a blank test, adding the ferrocyanide one drop at a time and making a test after each drop. In the course of a minute a brown tinge will show in all the tests except the first. Multiply the number of cubic centimeters used by the percentage value of 1 cc. to obtain the per cent. of zinc in the ore.

4. Ordinarily, any sulphides precipitated by the hydrogen sulphide water need not be filtered off, as the discoloration produced by even 10 per cent. of copper, for instance, does not too badly mask the uranium test. With a large amount of copper present the procedure may be as follows: After adding an excess of 3 cc. of strong hydrochloric acid to the filtrate and diluting it to about 200 cc. with hot water, pass in hydrogen sulphide gas to saturation. Filter through an 11-cm. filter and wash the precipitate thoroughly with hot water. If a little copper oxidizes and runs into the filtrate it will do no harm, as it is immediately reprecipitated as sulphide. Heat the filtrate, which should have a final volume of about 250 cc., nearly to boiling, and titrate as usual.

5. **Notes.**—*a.* It is not sufficient to wash the residue of ferric hydroxide, etc., simply with hot water. An appreciable amount of zinc is likely to be retained unless the residue is thoroughly washed with the hot ammoniacal solution prescribed. A direct test with a 40 per cent. zinc ore showed a retention of 1.27 per cent., which was afterward easily washed out with the ammoniacal solution.

b. The use of granulated test lead as a precipitant for copper is not recommended. Lead acts more slowly than hydrogen sulphide, sometimes fails to entirely remove the copper, and, unless the precaution of a special procedure be adopted, will always produce a solution with an unknown and variable degree of acidity, thus influencing the end-point. For a method using test lead see **8.**

c. Metallic aluminum appears to be entirely unsuitable for precipitating the copper, unless the aluminum which goes into solution is subsequently removed. Aluminum salts affect the titration and produce incorrect results. In carefully conducted experiments I have always found this to be the case, under my usual working conditions.

d. Even when the zinc contents of a solution are approximately known, it is not advisable, in the titration, to omit working with a reserve portion and proceeding in the regular manner. The reaction between the ferrocyanide and the zinc may be varied, at least temporarily, by varying the conditions, and a false end-point may be obtained by working too rapidly.

e. Cadmium is not completely precipitated by hydrogen sulphide from a hydrochloric acid solution containing considerable alkali chloride. In the regular method (and likewise the alternate and test lead methods), therefore, cadmium is to a greater or less extent counted as zinc, small amounts not being precipitated at all by the hydrogen sulphide. Theoretically, 100 Cd=87.3 Zn. (Actually, about 62 Zn.—Waring.)

f. Arsenic, when present in large amount, may make trouble by retaining iron in the ammoniacal solution filtered from the residue. No attention need be paid to arsenic unless its presence in excess is thus indicated. In such a case begin anew and give the ore the following preliminary treatment: Dissolve the ore in the usual way, but use an 8-oz. copper flask instead of an Erlenmeyer. Boil the solution nearly to dryness, then add 10 cc. of strong hydrochloric acid and boil until it is half gone (to expel nitric acid). Now add about 4 cc. of a solution of 1 gram of sulphur in 5 cc. of bromine and boil gently for a few moments. Then add 4 cc. of strong sulphuric acid and boil, by manipulating the flask over a free flame, until it is nearly all driven off. The arsenic will thus be sufficiently expelled. Cool, add 10 grams of ammonium chloride and 20 cc. of water and finish the assay as described in 6.

g. The opinion seems to be very common that gelatinous silica, unless well dehydrated, is liable to combine with and retain some of the zinc on the addition of the ammoniacal extraction solution. My own experiments have invariably indicated that this is not the case. A large amount of gelatinous silica may hold some zinc mechanically and thus render its complete extraction and the thorough washing of the residue more difficult. The partial dehydration obtained by following the regular method is usually quite sufficient to overcome this trouble.

6. **Alternate Method.**—I have observed that in the presence of a large excess of ammonium chloride, ammonia will precipitate iron entirely free from zinc. This fact may be applied to the assay of an ore as follows:

Treat 0.5 gram in an 8-oz. copper flask with 5 cc. of strong hydrochloric acid and 10 cc. of strong nitric acid and boil to pastiness. The final use of 5 cc. of strong sulphuric acid is advisable if gelatinous silica separates. In this case boil until the sulphuric acid is nearly gone—best by manipulating the flask over a free flame. Cool sufficiently, add 10 grams of ammonium chloride and 20 cc. of water and boil the mixture to effect solution of everything soluble. If basic salts remain, add a few drops of hydrochloric acid to dissolve them. When the solution is clear, remove from the heat, add strong ammonia water in about 5 cc. excess, then 20 cc. of saturated bromine water and boil for a few minutes. If the blackish brown color of the precipitate indicates much manganese, add more bromine water to insure its complete precipitation, and again boil. In doubtful cases test the filtrate with bromine water. Filter and finish as in the regular method (3).

7. Particular pains must be taken to remove all manganese with the bromine water, since none is separated during the decomposition, as in the regular method. When the amount

of manganese present is in excess of a few per cent. it is not safe to precipitate it with bromine, since it is liable to carry down an appreciable quantity of zinc as manganite (XVIII, 12). In such a case I would not advise the use of the alternate method.

8. Method Using Test-lead to Precipitate Copper.*—Begin as in the regular method 3). Use hydrogen peroxide (6 cc. or more) instead of bromine water to precipitate the manganese. (Bromine and its compounds are more difficult to subsequently remove.) Boil for several minutes to expel the excess of peroxide. Filter and wash as usual. Make the filtrate very slightly acid with hydrochloric acid, using methyl orange as indicator. Add about 20 grams of granulated test-lead, cover the beaker, and boil until all the copper is precipitated, i.e., until the solution has become perfectly colorless and then a little longer. Dilute with hot water to about 250 cc., add 3 cc. of strong hydrochloric acid and 1 cc. of a 50 per cent. solution of sodium thiosulphate, and titrate as usual at an initial temperature of about 85° C., using the uranium indicator and making the proper deductions as in the regular method.

To standardize the ferrocyanide solution, dissolve about 0.200 gram of zinc in 10 cc. of 1:2 hydrochloric acid, add 7 grams of ammonium chloride, make alkaline with ammonia and then reacidify slightly with hydrochloric acid, using methyl orange as indicator. Add 3 grams of test-lead, heat to boiling, dilute to about 250 cc. with hot water, add 3 cc. of strong hydrochloric acid and 1 cc. of a 50 per cent. solution of sodium thiosulphate and titrate as just described.

9. Treatment of Refractory Ores, etc.—No exact line of treatment can be prescribed for this class of material. The following will usually suffice: Begin according to the regular

* Method used by Mr. Chas. S. Curtis, U. S. Assayer, in the U. S. Customs Assay Office, Kansas City, Mo.

method (3). As soon as decomposition is as complete as possible, dilute somewhat and filter from the insoluble residue, washing the latter with hot water. Receive the filtrate in an 8-oz. Erlenmeyer flask and allow it to concentrate considerably while treating the residue. If the residue is likely to contain lead, wash it next with the hot sodium acetate solution used in the lead assay (XVI, 1), and then thoroughly with hot water to remove the sodium acetate, rejecting these filtrates. Ignite the washed residue at a low temperature in a platinum dish to remove the filter paper. Cool, add equal parts of strong hydrochloric and hydrofluoric acids, and digest on a water-bath until solution is complete, adding more acids if necessary. Finally, evaporate to small bulk and then add more hydrochloric acid and evaporate again to small bulk to remove most of the hydrofluoric acid. Wash this solution into the main portion which is concentrating. Boil the combined solutions to small bulk, add 10 cc. of strong nitric acid and boil nearly to dryness. Now add 12 cc. of strong nitric acid and 5 grams of potassium chlorate and continue as in the regular method.

Sometimes the ore or other material may be treated at the outset with hydrochloric and hydrofluoric acids, in platinum, as just described, for the residue.

The washed insoluble residue may also be fused with alkali carbonate, either alone or mixed with borax glass, the melt disintegrated with water, and the mixture transferred to the flask containing the main solution, finishing as above.

10. Modification to Avoid Cadmium.—If it be desired to avoid the influence of cadmium, either of the following methods may be adopted:

Method No. 1.—Begin by either the regular or alternate method and proceed as usual until the filtered ammoniacal solution is obtained. To this filtrate add a few cubic centimeters of strong ammonium sulphide solution. It is usually unnecessary to add more than is required to convert to sulphides all the cadmium and lead present. Stir the mixture, and then add 3 grams of potassium cyanide, either as the coarsely crushed solid or in strong solution. Stir the mixture until all the zinc sulphide has dissolved, allow to stand a short time for the cadmium to separate, and then filter. Copper goes into solution with the zinc, while cadmium and lead remain insoluble. As the former is liable to clog the filter badly, it is best to make use of Dittrich's paper-pulp idea * as follows: Place one-half of a 9-cm. filter in a 6-oz. flask together with a little hot water. Cork the flask and shake it violently until the paper is reduced to a fine pulp. Pour this mixture into an 11-cm. filter placed in a funnel and allow the water to drain off. Now place a 400-cc. beaker under the funnel and filter the cyanide solution through the paper pulp. Wash out the beaker with cold water, and then wash filter and precipitate with a cold 2 per cent. solution of potassium cyanide. Remove the filtrate to a hood, add some litmus solution as an indicator, and acidify with strong hydrochloric acid, finally adding 10 cc. in excess. If no potassium chlorate is present add 2–3 grams. Cover the beaker, and boil until the hydrocyanic acid is expelled and the mixture has clarified. Finally remove from the heat, again add some litmus solution, neutralize with ammonia, and then reacidify with 3 cc. excess of strong hydrochloric acid. Have the volume of the solution as near as possible to 200 cc. To the nearly boiling liquid add 50 cc. of strong hydrogen sulphide water and titrate as usual.

Method No. 2. — Begin as usual for the Alternate Method, finally boiling with 8 cc. of strong sulphuric acid until half of the latter is expelled — best by manipulating the flask over a free flame. Cool, add 25 cc. of water, heat to effect solution, and then

* I, 12.

add a piece of stout sheet aluminum, about $1\frac{3}{4}$ in. long and $\frac{5}{8}$ in. wide. Boil the mixture for perhaps 10 minutes, or until the solution is reduced and any copper present is precipitated. Remove from the heat, dilute, add litmus solution as indicator, and then make slightly alkaline with ammonia. Now make faintly acid with dilute sulphuric acid (1:4), and then add 5 cc. in excess. Add, now, 25 cc. of strong hydrogen sulphide water. This will precipitate any remaining traces of copper and also ensure the complete precipitation of the cadmium. Filter, washing with cold water and receive the filtrate in a 12-oz. flask. Boil the solution until the hydrogen sulphide is expelled, then add 5 cc. of strong nitric acid to oxidize the iron, and continue the boiling until the solution is reduced to a bulk of 50 cc. or less. Now add 10 grams of ammonium chloride, and when it has dissolved add ammonia in about 5 cc. excess, then 20 cc. of saturated bromine water and finish the assay as described for the alternate method (6). Do not omit the 50 cc. of hydrogen sulphide water at the end.

11. **Waring's Zinc Method.**—This method depends upon the separation of the zinc from manganese, iron, and aluminum by means of hydrogen sulphide, under slight pressure, in a solution very slightly acidified by formic acid; the metals of the copper group having been previously separated by metallic iron or aluminum with simultaneous reduction of ferric salts. The operations are:

(1) *Solution.*—The calamine, willimite, franklinite, blende, and other soluble minerals, or ores containing them, are decomposed by hydrochloric acid or aqua regia, with subsequent treatment and evaporation with an excess of hydrochloric or sulphuric acid to thoroughly eliminate nitrous compounds. If zinc spinels or aluminates are present, the insoluble residue must be fused with a mixture of sodium carbonate and borax glass, the fused mass dissolved and the solution added to the main one. If much

silica is present, spinels are decomposed by fusion with sodium carbonate in a platinum crucible, any lead sulphate present having been extracted by ammonium acetate. In the absence of silica or boric acid, the spinels cannot be decomposed by fusion with sodium carbonate alone. In such a case they can be decomposed by prolonged fusion with an alkali bisulphate. Silicates, such as cinders from oxide furnaces, unchilled slags, and some natural silicates undecomposable by acids, must be fluxed or cintered with sodium carbonate before treatment with hydrochloric acid. It is not necessary, in any case, to evaporate to dryness to separate silica—it can be filtered off in the gelatinous state. This can be done very rapidly, after dilution with water, when the gelatinization has reached a maximum and before dehydration has begun. The gelatinous silica at this stage will not retain any traces of metals after a few washings.

(2) *Reduction.*—To avoid the effects of reactions like

$$2FeCl_3 + H_2S = 2FeCl_2 + 2HCl + S,$$

and at the same time to remove copper, silver, and bismuth before precipitating with hydrogen sulphide, the filtered solution, made fairly acid with hydrochloric or sulphuric acid, is boiled for fifteen or twenty minutes with a strip of clean sheet iron or steel. By this treatment all the metals likely to be precipitated with zinc as sulphides are separated, except cadmium, which is not in the least degree reduced by metallic iron.

Mr. G. C. Stone has suggested the use of metallic aluminum for the reduction. This has the advantage of separating cadmium and lead along with the other metals of the copper group, since both are completely precipitated by aluminum from a rather strongly acid boiling solution of sulphates or chlorides, so that when zinc only is to be determined, the subsequent operations are very much shortened.

Reduction may also be effected by means of sodium sulphite

or thiosulphate, when copper, or copper and aluminum are to be determined from the same weighed portion.

The reduction is followed by filtration, the filtrate being received in a flask of about 300-cc. capacity.

(3) *Neutralization.*—Add to the filtrates a drop of methyl orange, then run in, from a pipette, a rather dilute solution of sodium hydroxide, meanwhile constantly agitating the contents of the flask with a swirling motion, until the pink color barely, but permanently, changes to a light yellowish tint and the cloudiness, due to the separation of hydroxides, fails to clear up entirely. Then add, drop by drop, enough 50 per cent. formic acid (sp. gr. 1.12) to just restore the permanent pink color, and add up to half a cubic centimeter additional. Dilute the solution to 200 or 250 cc. (or so that it will contain not more than 0.15 to 0.20 gram of metallic zinc in 100 cc.) and heat to about 80°.

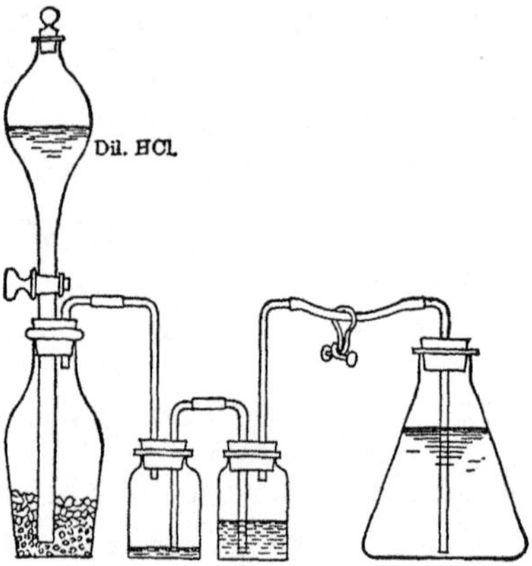

FIG. 20.

(4) *Precipitation.*—A rubber stopper, through which passes the delivery-tube from a source of supply of hydrogen sulphide,*

* The apparatus used by Waring is shown in Fig. 20. It is adapted from that of C. A. Brewer. The lower end of separatory-funnel is somewhat contracted and extends to bottom of coarse shot upon which rests the FeS.

is loosely placed in the neck of the flask and a moderately rapid stream of gas allowed to pass through the hot liquid. When the precipitation of zinc sulphide is well under way, the stopper is pushed in tightly. Absorption of the gas ceases when all the zinc is precipitated; the precipitate settles quickly, and the gas pressure rises rapidly when the operation is completed. When several precipitations are to be made at the same time, the flasks are arranged in succession in the usual manner and the first is removed when the precipitation is well started in the third, and so on, changing the gas connections as required. The outlet from the last flask is not closed until the precipitation is partially completed therein. Numerous experiments have shown that zinc can be completely precipitated and separated from iron, manganese, and alumina under the conditions named, by the passage of only a very little more than the amount of hydrogen sulphide theoretically required. The use of a large excess is therefore unnecessary and is also undesirable.

(5) *Treatment of the Precipitate.*—When the preceding operations have been properly performed, the precipitated zinc sulphide will be pure white, pulverulent, and very easily filtered and washed. Hot water only need be used for washing, no zinc will dissolve or pass through the filter, as is the case with the slimy zinc hydrosulphide precipitated from cold solutions in the usual manner. Pour the contents of the flask upon a filter at once and wash with hot water. Spread the filter with its contents upon a large watch-glass or on the inner wall of a capacious beaker, and wash the precipitate into the bottom of the beaker by a jet of hot water. Wash the precipitating-flask and the lower end of the gas delivery-tube with 10 cc. of strong hydrochloric acid, followed by hot water, pouring the acid and washings successively over the washed filter on to the precipitate in the beaker. When the volume of the acid solution has reached

125 to 130 cc. the solution is warmed gently to dissolve the zinc sulphide completely.

When cadmium is also present, (i.e., when the reduction has been effected by metallic iron), the zinc sulphide will dissolve completely before any cadmium sulphide is dissolved. By practice and experience, the point when all the zinc is dissolved and only the brilliant yellow cadmium sulphide remains, can easily be distinguished. The solution is then further diluted with an equal volume of hydrogen sulphide water, allowed to settle and the cadmium sulphide filtered off and estimated in the usual way by warming with acid ferric sulphate and titrating with permanganate. The iron equivalent of the permanganate used multiplied by 1.006 equals the cadmium. Properly performed, the result is accurate * (Waring).

The solution of zinc sulphide in dilute hydrochloric acid is heated to 60° or more, made up to 200 or 250 cc. with hot water, a little ammonium chloride added, and it is ready for titration with ferrocyanide (see 2).

The method of separating zinc by hydrogen sulphide from nickel, cobalt, and manganese, recommended by Fresenius,† is not applicable when iron is present, as iron is thrown down by hydrogen sulphide in the presence of sodium acetate.

12. Modification. For Low-grade Zinc Ores, Slags, Furnace Residues, etc., and for all purposes wherein it is required to determine small amounts of zinc accurately.

Proceed exactly according to the method just described until the zinc sulphide precipitate has been washed, then, instead of dissolving the precipitate, dry and ignite it carefully in a clean

* Treadwell says (Anal. Chem., Hall, 2d Ed., II, p. 167): "It is not possible to precipitate pure cadmium sulphide from acid solutions by means of hydrogen sulphide; the precipitate is always contaminated with a basic salt (Cd_2Cl_2S — Cd_2SO_4S, etc.), whether the precipitation takes place in cold or hot solutions, whether under atmospheric pressure or under increased pressure (in a pressure-flask), and in fact the amount of basic salt formed increases with the amount of free acid present."

† American edition, sec. 160, paragraphs 74 and 75.

muffle without separating it from the paper. No loss of zinc will occur, nor will basic sulphate be formed, if the wet precipitate is ignited at the mouth of the muffle until the paper is consumed, and the oxidation of the residue is then conducted at a low temperature (about 450°) until at the last, when it may be moved back to where the temperature is near the melting-point of silver.

As much as 0.15 gram of zinc sulphide can be completely oxidized in this way in forty to sixty minutes. The calcination may be effected in a smooth shallow scorifier, an inch and a half in diameter, from which the calcined oxide can be brushed into the scale-pan without appreciable loss. Calculate the zinc from the ZnO found by multiplying by 0.8035.

Waring's method has been modified both by its author and by Dr. H. C. P. Weber of the U. S. Bureau of Standards, and the following scheme including the changes is proposed by the Zinc Committee of the American Chemical Society.

13. **Modified Waring Method.*** — After decomposing the weighed sample by acids alone, or aided by fusion, as the case may require, all the zinc is to be brought into solution as sulphate. If nitric acid was used in the decomposition, all traces of it must be expelled by evaporation with hydrochloric and sulphuric acids successively, or by two evaporations with sulphuric acid, finally to abundant evolution of SO_3 fumes. Dissolve the mass in 25 to 40 cc. of water and add sufficient sulphuric acid to bring the free acid in the solution up to 10 or 15 per cent. Introduce piece of heavy sheet aluminum and boil 10 minutes, or to complete reduction. Filter, and wash through a filter containing a piece of aluminum into a beaker containing a stirring-rod or strip of the same metal, cool, add a drop of methyl orange, and neutralize carefully with *sodium bicarbonate* to a light straw color. Add, dropwise, dilute formic acid (20 per cent. strength) until

* Jour. Am. Chem. Soc., XXIX, 265.

the pink color is just restored, then 5 drops more. (Dilute hydrochloric acid, 1 part strong acid to 6 parts water, may be substituted for formic acid when ammonium thiocyanate* is to be introduced.) Dilute to about 100 cc. for each 0.1 gram of zinc possibly present, add, if much iron is present, 2 to 4 grams ammonium thiocyanate, remove the strip of aluminum, heat nearly to boiling, and saturate with hydrogen sulphide. Allow the pure white zinc sulphide to subside for a few minutes, then filter and wash with hot water. Transfer precipitate and filter to a capacious beaker, heat with 8 or 10 cc. of strong hydrochloric acid and 30 or 40 cc. of water, until the zinc is all in solution. Determine the zinc as pyrophosphate containing 42.91 per cent. zinc, or by titration with ferrocyanide. The use of ammonium heptamolybdate† in one per cent. solution as an indicator, instead of uranium acetate or nitrate, is recommended, provided all free hydrogen sulphide has been previously expelled from the solution by heating. If a blue color still appears in the test drop, add a crystal or two of sodium sulphite to the zinc solution, to decompose any remaining hydrogen sulphide.

14. **Notes.** — Using the thiocyanate modification and the ferrocyanide titration, I have found the above very accurate, except that cadmium was not entirely removed — at least in presence of copper. To conform to my method of standardization proceed as follows: Drop filter with zinc sulphide into a 6-oz. flask. Cleanse beaker and delivery tube with 25 cc. of hydrochloric acid and rinse into the flask. Boil to dissolve the zinc, disintegrate the filter and expel all hydrogen sulphide. Add a little litmus solution or paper, neutralize with ammonia, and then reacidify with 3 cc. excess of hydrochloric acid. Transfer to a 400-cc. beaker, dilute to 250 cc. and titrate hot, as usual.

* Zimmermann, Ann. Chem., (Liebig), 99, 1.

† Nissenson and Kittembeil, Chem. Zeit., 77 (1905), 951–955.

15. Notes by Waring and Stone.—Under the conditions described the precipitation of cadmium is complete, but traces of copper remain in solution unless the boiling is continued for a very long time; this prolonged boiling is not necessary as the copper is afterwards precipitated as sulphide with the zinc and does not redissolve with the latter.

In neutralizing, if the solution is strongly acid, it is better to nearly neutralize with sodium or potassium hydroxide and finish with bicarbonate. This not only saves time but lessens the chance of loss by foaming.

It is not necessary to pass the hydrogen sulphide under pressure if the solution is diluted as already directed. The gas should be passed through the solution until a drop of the liquid blackens a drop of cobalt or nickel sulphate or chloride made alkaline with ammonia. After a little experience it is not necessary to make even this test. It is very important that the zinc solution should be quite hot during the precipitation of the sulphide, therefore it is advisable to begin nearly at the boiling-point and to pass the gas rapidly. If the heating of the solution has taken much time, the excess of formic acid may volatilize (if this reagent has been used). In such cases enough more must be added to make the solution acid.

We most strongly recommend that zinc be determined gravimetrically by weighing as pyrophosphate, except where the operator has had much and recent practice with the ferrocyanide titration. While the latter is capable of giving very accurate results it will only do so when *all* the conditions are *exactly* the same both in standardizing and during the final titration, and it is necessary to have considerable experience with the method to be sure of accuracy.

16. Weighing the Zinc as Pyrophosphate.* — Filter the

* Method of Geo. C. Stone. W. Geo. Waring, Jour. Am. Chem. Soc., XXVI, 28.

solution of the zinc sulphide in hydrochloric acid finally obtained in Waring's Method and bring the filtrate to a cold, dilute and slightly acid condition. Now add a *large* excess of ammonium sodium hydrogen phosphate and then neutralize very carefully with ammonia, adding it drop by drop, finally adding a drop or two in excess. Finally, add about 1 cc. of acetic acid and warm gently until the flocculent precipitate of $ZnNH_4PO_4 + H_2O$ has settled completely as a dense crystalline powder. Filter and wash with hot water.

Dry the filter and precipitate, separate the paper and burn it, add the ash to the residue and ignite the two, gently at first, then for a few minutes at a bright red heat. Cool and weigh as $Zn_2P_2O_7$, containing 42.77 per cent. of zinc.

The flocculent $ZnNH_4PO_4 + H_2O$ is very soluble in the mineral acids as well as in ammonia, but after crystallization it is much less soluble in the latter. It is only slightly soluble in acetic acid; an excess of 1 cc. in 100 cc. of solution does not dissolve an appreciable quantity. It is somewhat soluble in all ammonium salts, if only a small excess of phosphate is present. The addition of 1 cc. of a 10 per cent. solution of sodium ammonium phosphate for each 0.005 gram of zinc is sufficient to entirely prevent its solution in ammonium chloride or sulphate or in the acetate unless the latter is present in enormous quantity. It is, however, always slightly soluble in the oxalate. Therefore, for very accurate work, lime and magnesia, if present in the zinc solution, are preferably separated together as phosphates, after adding a large excess of ammonia and reprecipitating; then the combined filtrates are to be slightly acidulated and proceeded with as above. The crystalline zinc ammonium phosphate is quite insoluble in hot or cold water.

CHAPTER XXXI.

COMBINING DETERMINATIONS.

IT is an ordinary occurrence for the technical chemist to have to determine several constituents in the same sample of ore. The expert operator will, of course, be able to combine his regular methods for single constituents, whenever possible or convenient, so as to make two or more determinations from the same weighed portion of ore, and thus save considerable time. The accuracy of the work is sometimes lessened by such a combination, and its adoption will then depend upon the rigor of the requirements.

The following examples are perhaps sufficient to show how work can be hastened in a busy laboratory, using only technical methods. Of course, by the tedious operations of exact analysis, many constituents can sometimes be determined from the same weighed portion, but such methods are not under consideration. Where much extra manipulation is involved, it is usually not worth while to work by a combination method.

1. **Copper, Lead and Insoluble.**—Make the copper as usual (XIII, 3), but take the precaution to cool the solution before the first filtration and to wash the lead sulphate residue with dilute sulphuric acid. Dissolve the lead sulphate on the filter with the usual hot sodium acetate solution and proceed with

the filtrate as described for lead (XVI, 4). Thoroughly wash the residue remaining on the filter with hot water, to remove the sodium acetate, and then ignite filter and contents and weigh as insoluble residue.

2. Zinc, Iron and Insoluble.—Make the zinc as usual (XXX, 3). The iron and insoluble residue will be either on the filter or adhering in the flask. Rinse the precipitate on the filter into a beaker with hot water, using as little as possible. Add 5 cc. of strong hydrochloric acid and warm the mixture until the ferric hydroxide, etc., has dissolved. Transfer this solution, and any insoluble residue, into the original flask and again warm, if necessary, to dissolve any adhering precipitate. Now filter through the same filter as before, thus dissolving any precipitate thereon, receiving the filtrate in an 8-oz. flask. Wash flask and filter with hot water acidulated with a few drops of hydrochloric acid, and see that all the residue is transferred from the flask to the filter, using a rubber-tipped bent glass rod, if necessary, to detach adhering particles. Concentrate the filtrate by boiling, if too bulky, and determine the iron in it as usual (XV, 31).

The insoluble residue is now ready to ignite unless it is liable to contain lead. In this case dissolve out the lead with a strong hot solution of ammonium chloride, acidulated with hydrochloric acid, or employ the hot sodium acetate solution used in the lead assay (XVI, 1). Finally, wash thoroughly with hot water, then ignite filter and contents and weigh as usual.

4. Calcium and Magnesium.—It is evident that no modifications of the methods described are necessary in this case. Simply proceed as for CaO (X, 1), and in the filtrate from the calcium oxalate determine the MgO in the usual manner (XVII, 1, 2).

5. Insoluble, Iron, Calcium and Magnesium.—Proceed as for insoluble residue by any appropriate method (XXIV, 2). Wash the filtered residue with hot water slightly acidulated with hydrochloric acid and then treat it as in 3 above. If it requires purification from lead, do not add the resulting filtrate to the original filtrate.

Proceed with the original filtrate as for calcium (x, 1), and thus obtain the iron as ferric hydroxide, the calcium and magnesium being in the filtrate (or filtrates). Determine the latter as above (4).

If the ferric hydroxide is small in amount, dissolve it on the filter by pouring over it a hot mixture of 10 cc. of strong hydrochloric acid and 20 cc. of water, receiving the filtrate in an 8-oz. flask. It is best to wash larger amounts into a beaker with a little hot water, add 5–10 cc. of hydrochloric acid and warm until the iron precipitate has dissolved, pouring the solution through the filter to dissolve what remains there. Wash the filter with hot water slightly acidulated with hydrochloric acid. Concentrate the filtrate by boiling, if necessary, and determine the iron in it in the usual manner (xv, 31).

6. Copper and Iron.—Proceed as for copper by the iodide method (xiii, 3), determining the copper as usual. The filtrate from the precipitated copper will contain the iron all reduced, but also containing hydrogen sulphide. Boil the solution in a flask until the hydrogen sulphide is completely expelled, then transfer it to a battery-jar containing some cold water, dilute to 700 cc. with cold water, add 5 cc. of strong sulphuric acid and titrate with permanganate, as described in xv, 8. Of course the aluminum used should be practically free from iron.

It will be observed that the above combinations can be combined with each other in certain cases, for instance,

7. Insoluble, Lead, Copper and Iron.—Proceed as in 1, determining the lead and insoluble in the residue, and then determine the copper and iron in the filtrate according to 6.

CHAPTER XXXII.

BOILER WATER.

FULL directions for the analysis of water to be used for steam purposes can be found in most of the books on technical analysis, and it is not intended to describe an elaborate method in this work. The following scheme can be carried out very quickly, and is sufficiently accurate in most cases:

If the water is turbid, allow it to settle or filter it so as to remove the suspended matter.

1. **Total Solids.**—Evaporate 100 cc. in a weighed platinum dish on the water-bath to dryness. If the dish is too small to hold 100 cc. at once, measure the water into a beaker and transfer it to the dish as the evaporation proceeds. A 100-cc. pipette is most convenient for the measuring, as no washing out is necessary and there is, therefore, no wash-water to evaporate. After evaporation, place the dish and residue in an air-bath and heat at 150° C. for thirty minutes. Cool in desiccator and weigh.

2. **Organic and Volatile Matter.**—Hold the dish containing the dry residue in the tongs and manipulate it over a Bunsen flame so as to burn off the organic matter at as low a temperature as possible. Sometimes the burned residue remains dark owing to the presence of manganese. If this is mistaken for unburned carbon too high a temperature may be employed in

attempting to consume it and more or less carbon dioxide may be expelled from carbonates in consequence. This carbon dioxide can be replaced by adding a little CO_2 water and evaporating to dryness, but the extra trouble is hardly worth while, as the possible errors of this determination are liable to make it unsatisfactory in any event. The result obtained is seldom used, but it sometimes proves useful as a check. After the ignition cool the dish in a desiccator and weigh, noting the loss from the last weight.

3. **Conversion to Sulphates.**—Add sufficient dilute sulphuric acid to the last residue in the dish (a few drops are usually enough) and turn the latter in various positions so that the acid will come in contact with all the residue. Note whether or not an effervescence indicating carbonates is produced, as knowledge on this point may prove useful later. Heat the dish cautiously so as to expel the water and free sulphuric acid and then raise the temperature sufficiently to decompose any iron sulphate and leave ferric oxide; avoid unnecessary overheating however. Cool and weigh as combined Na_2SO_4, $CaSO_4$, $MgSO_4$, Fe_2O_3, and SiO_2. This weight will be used in the calculation of results.

4. **Silica.**—Warm the sulphate residue with sufficient dilute hydrochloric acid to dissolve all the sulphates. If there is much ferric oxide it is best to warm with a little strong acid at first until it has dissolved. Filter through a small ashless filter and wash with hot water. Sometimes it is a good plan to loosen any insoluble matter on the sides of the dish with a rubber-tipped glass rod and continue the heating with dilute acid to make sure that all sulphates are dissolved. Return the filter and any insoluble residue to the dish and ignite. After cooling, weigh as SiO_2.

5. **Ferric Oxide, etc.**—Heat the filtrate from the silica nearly to boiling and add a slight excess of ammonia, preceded, if manganese is possibly present, by 5 cc. of strong bromine water.

Bring to a boil and then allow to stand over a low flame for a short time for the precipitate to collect. Finally, filter through a small ashless filter, wash with hot water, ignite and weigh, after cooling, as Fe_2O_3. The amount is usually so small that the platinum or porcelain crucible used for the ignition need not be weighed, but the ferric oxide transferred directly to the scale-pan. Only in rare cases where Mn_3O_4 or Al_2O_3 are present in unusual amount is it necessary to make any special separation of these substances.

6. **Calcium Oxide.**—Heat the filtrate from the ferric hydroxide to boiling, add sufficient ammonium oxalate solution and allow to stand, hot, till the precipitate has settled. Filter, wash well with hot water, and determine the CaO in the calcium oxalate by titration with potassium permanganate, as in X, 4.

7. **Magnesium Oxide.**—Concentrate the filtrate from the calcium oxalate to smaller bulk if necessary, and then precipitate the magnesium as magnesium ammonium phosphate, as described in XVII, 2, finally weighing as $Mg_2P_2O_7$ and calculating to MgO.

8. **Alkalies.**—For the purposes of this analysis it is usually sufficient to regard the alkalies as consisting wholly of Na_2O. The amount of Na_2O is determined by calculation as follows: From the weight of the combined sulphates, etc., as determined in 3, subtract the calculated weights of the calcium and magnesium sulphates, and also the Fe_2O_3 and SiO_2. Consider the remainder as Na_2SO_4 and calculate from it the corresponding weight of Na_2O.

For use in these calculations the table on page 317 is appended.

9. **Sulphur Trioxide.**—Slightly acidify 100 cc. of the water with hydrochloric acid, heat to boiling and slowly add an excess of barium chloride solution. Continue as described in XV, 1, finally weighing the precipitated $BaSO_4$, from which the SO_3 may be calculated. $BaSO_4 \times 0.343 = SO_3$.

10. **Chlorine.**—Measure 100 cc. of the water into a clean porcelain dish or casserole and determine the chlorine volumetrically by Mohr's method, described in XI, 1. Instead of a N/10 solution of silver nitrate, it is better to prepare a solution containing 4.7936 grams of $AgNO_3$ per liter. 1 cc. of this solution will equal 0.001 gram of chlorine.

11. By the scheme as described, three portions of water of 100 cc. each have been required in making all the necessary determinations. Considerable time may be saved, with perhaps some loss of accuracy, by using a separate portion for determining the Fe_2O_3, CaO, and MgO, instead of waiting for the residue obtained by evaporation. The error in this plan will be mainly due to the failure to remove silica, thus, possibly, causing slight increases in the results for Fe_2O_3 and MgO. In most cases, however, this error is so trifling as to be negligible. When this method is adopted do not fail to acidify the water at the outset with 5 cc. of strong hydrochloric acid, so that the subsequent addition of ammonia will form sufficient ammonium chloride to prevent the precipitation of any calcium as carbonate.

12. **Calculation of Results.**—In combining the acids and bases only an arbitrary rule can be followed, and even this is subject to modification in special cases according to the judgment of the analyst. In general, proceed as follows:

Combine the chlorine with sodium. If there is an excess of chlorine combine it with calcium, then with magnesium.

If there is an excess of sodium, combine it as Na_2O with SO_3. If the SO_3 is insufficient for all the available Na_2O, calculate the excess of the latter to Na_2CO_3.

If SO_3 is in excess combine it with CaO and then with MgO. Calculate any excess of CaO or MgO to $CaCO_3$ and $MgCO_3$.

Calculate the Fe_2O_3 to $FeSO_4$ if any SO_3 remains after satisfying the other bases, otherwise to $FeCO_3$. It will be observed that the iron is regarded as being in a ferrous condition. This is due to the presence of organic matter usually. If of sufficient importance the proper qualitative tests may be made to determine the condition of the iron.

The silica is usually reported as SiO_2, although in water containing much sodium carbonate it may exist as sodium silicate. I have observed cases where such a combination was apparently necessary to satisfy the conditions.

It will be observed that the amount of carbon dioxide is obtained entirely by calculation. In the majority of cases this will suffice, as a fairly correct idea of the general composition of the dissolved solids is all that is required. The presence of a large amount of calcium or magnesium carbonate usually indicates the existence of free carbon dioxide, which probably forms soluble acid carbonates of the alkaline earths.

13. In this analysis it is sufficiently accurate to regard milligrams per liter as equivalent to parts per million. As each determination is made on the basis of 1/10 of a liter, the results in milligrams, multiplied by 10 will give the parts per million. Therefore, after each result is obtained in grams, record it in decimilligrams or parts per million by moving the decimal point four places to the right. When the entire analysis has been calculated and tabulated in this way, subtract the sum of the constituents, except organic and volatile matter, from the total solids and call the difference organic and volatile matter. It will usually be less than the figure as actually determined for the reasons stated above.

Finally, for technical purposes, it is customary to report the results in grains per U. S. gallon of 231 cubic inches instead of in parts per million. The conversion is easily made by means of the table on page 316.

14. Example of Calculating Analysis of Boiler-water.—Let us suppose the various determinations have resulted as follows, in parts per million:

Total solids	928.0
Organic and volatile matter	179.0
Sulphates, etc.	934.0
SiO_2	12.0
Fe_2O_3	4.0
CaO	165.0
MgO	44.7
SO_3	246.5
Cl	60.0

All the constituents are shown except Na_2O, which is determined as follows: Calculate the CaO and MgO to sulphates by multiplying by the proper factors given in the table, and then add together the following:

$CaSO_4$	400.4
$MgSO_4$	133.4
Fe_2O_3	4.0
SiO_2	12.0
	549.8

Subtract this sum, 549.8, from the sum of the sulphates, etc., 934.0, and consider the remainder, 384.2, as Na_2SO_4. This, using the table, is equivalent to 167.8 Na_2O.

Now combine the Cl with Na. 60 Cl = **99.0 NaCl** (omitting more than one decimal place). 99.0 − 60 = the Na required, or 39.0. This is equivalent to 52.5 Na_2O. Subtracting this from the original 167.8 Na_2O there remains an excess of 115.3 Na_2O.

Combine this with SO_3. 115.3 Na_2O = **263.9 Na_2SO_4.** 263.9 − 115.3 = 148.6, the SO_3 used. Subtracting this from the total SO_3, 246.5, there remains an excess of 97.9 SO_3.

Combine this with CaO. 97.9 SO_3 = **166.5 $CaSO_4$.** 166.5 − 97.9 = 68.6, the CaO used. Subtracting this from the total CaO, 165.0, there remains 96.4 CaO excess.

Combine this with CO_2. 96.4 CaO = **172.0 $CaCO_3$.**

Calculate the remaining bases as carbonates.

44.7 MgO = **93.4 $MgCO_3$.** 4.0 Fe_2O_3 = **5.8 $FeCO_3$.**

The sum of the above results in bold type, together with the **SiO_2, 12,** = 812.6. Subtracting this from the *Total Solids*, 928.0, there remains 115.4 for the *Organic and Volatile Matter*. It will be noted that this is 68.6 less than the figure actually found after ignition, indicating that some CO_2 was driven off by the heat employed.

15. The entire analysis may now be tabulated as follows:

	Parts per Million.	Grains per Gallon.
Sodium chloride	99.0	5.78
Sodium sulphate	263.9	15.39
Calcium sulphate	166.5	9.71
Calcium carbonate	172.0	10.03
Magnesium carbonate	93.4	5.45
Ferrous carbonate	5.8	0.34
Silica	12.0	0.70
Organic and volatile matter	115.4	6.73
Total solids	928.0	54.13

The results in grains per gallon are the ones usually reported. The conversion from parts per million is quickly made by means of the table on page 316.

16. Valuation of Water from Results of Analysis. — The chemist is frequently required to give an opinion as to the value of a water for steam purposes, basing his judgment on his analysis. In this regard the following points may be considered: The principal scale-forming ingredient is calcium sulphate. Calcium and magnesium carbonates also form scale to a certain extent, especially if much calcium sulphate is present to act as a cementing agent. Much of the calcium and magnesium carbonates, held in solution by excess of carbon dioxide as acid carbonates, will tend to deposit as a sludge or mud rather than as scale when the water is boiled.

While the other constituents ordinarily found in water are not classed as scale-forming, yet they are practically all found to a greater or less extent in the scale on analysis, showing that with a sufficient cementing agent they all contribute to the trouble.

A large amount of alkali carbonate in water may cause foaming or priming in the boiler, as may also much organic matter.

Free acid is, of course, corrosive and is often found in mine waters owing to the oxidation of pyrites.

Magnesium chloride is regarded as an active corrosive agent, upon the supposition that it is decomposed in the boiler with the liberation of free hydrochloric acid. When more chlorine is present in a water (containing magnesium) than can combine with the sodium (and potassium), magnesium chloride is usually considered present.

The Union Pacific Railroad Company classifies the constituents of boiler water as follows:

Incrusting, or Scale-forming Solids:

Oxides of Iron and Aluminum,
Calcium Carbonate,
Calcium Sulphate,
Calcium Chloride (corrosive),
Magnesium Carbonate,
Magnesium Sulphate (forms scale only in the presence of calcium carbonate),
Magnesium Chloride (corrosive).

Non-incrusting Solids:

Alkali Carbonates, Sulphates and Chlorides, and Organic Matter.

The quality of a boiler water is judged as follows:

Foaming and Priming:

The amount of non-incrusting solids that a boiler water can contain before foaming or priming will take place is about 175 grains per gallon. The less non-incrusting solids that a given water contains, the more of the water may be evaporated before this degree of concentration will be reached.

The maximum limit permissible is generally taken as 50 grains per gallon, although this would be too high unless the engine occasionally took water very much lower in non-incrusting solids. For example, in the case of an average engine, of which the boiler contains 2000 gallons and the tender 6000 gallons, after the first tender of water was used, the non-incrusting solids in the boiler would have attained a concentration of 200 grains per gallon. Therefore, in judging the foaming or priming qualities of a boiler water, the service that the engine is performing, as well as the amount of non-incrusting solids in solution, must be considered.

Scale-forming Matter:

A water containing less than	15 grains per gallon	Good
" " " " "	15–20 grains per gallon	Fair
" " " " "	20–30 " " "	Poor
" " " " "	30–40 " " "	Bad
" " " " "	over 40 " " "	Very bad

CONVERSION OF "MILLIGRAMS PER KILOGRAM" INTO "GRAINS PER U. S. GALLON" OF 231 CUBIC INCHES.

Parts per Million.	Grains per Gallon.	Parts per Million.	Grains per Gallon.	Parts per Million.	Grains per Gallon.
1	0.0583	36	2.1001	71	4.1418
2	0.1167	37	2.1584	72	4.2001
3	0.1750	38	2.2167	73	4.2584
4	0.2333	39	2.2751	74	4.3168
5	0.2917	40	2.3334	75	4.3751
6	0 3500	41	2.3917	76	4.4334
7	0 4083	42	2.4501	77	4.4918
8	0.4667	43	2.5084	78	4.5501
9	0 5250	44	2.5667	79	4.6085
10	0.5833	45	2.6251	80	4.6668
11	0.6417	46	2.6834	81	4 7251
12	0.7000	47	2.7417	82	4.7835
13	0 7583	48	2.8001	83	4 8418
14	0 8167	49	2.8584	84	4 9001
15	0 8750	50	2.9167	85	4.9585
16	0 9333	51	2.9751	86	5 0168
17	0.9917	52	3.0334	87	5.0751
18	1.0500	53	3.0917	88	5.1335
19	1.1084	54	3.1501	89	5.1918
20	1.1667	55	3.2084	90	5.2501
21	1.2250	56	3.2667	91	5.3085
22	1.2834	57	3.3251	92	5.3668
23	1.3417	58	3.3834	93	5.4251
24	1.4000	59	3.4418	94	5.4835
25	1.4584	60	3.5001	95	5.5418
26	1.5167	61	3.5584	96	5.6001
27	1.5750	62	3.6168	97	5.6585
28	1.6334	63	3 6751	98	5.7168
29	1.6917	64	3.7334	99	5.7751
30	1.7500	65	3.7918	100	5.8335
31	1.8084	66	3.8501		
32	1.8667	67	3.9084		
33	1.9250	68	3.9668		
34	1.9834	69	4.0251		
35	2.0417	70	4.0834		

One U. S. gallon of pure water at 60° F., weighed in air at 60° F., at atmospheric pressure of 30 inches of mercury, weighs 58334.9+ grains. (Mason, Examination of Water.)

TABLE OF FACTORS FOR USE IN WATER ANALYSIS.

Given.	Required.	Factor.	Log.	Given.	Required.	Factor.	Log.
Ca	CaO	1.3990	0.1459	Na	NaCl	2.5390	0.4046
$CaCl_2$	CaO	0.5055	9.7037	Na	Na_2CO_3	2.3010	0.3620
CaO	$CaCO_3$	1.7840	0.2514	Na	Na_2O	1.3470	0.1294
CaO	$CaSO_4$	2.4270	0.3850	NaCl	Cl	0.6059	9.7824
$CaSO_4$	Ca	0.2945	9.4691	NaCl	Na	0.3940	9.5954
Cl	$CaCl_2$	1.5660	0.1947	NaCl	Na_2O	0.5308	9.7249
Cl	KCl	2.1040	0.3231	Na_2CO_3	Na_2SO_4	1.0640	0.1271
Cl	NaCl	1.6500	0.2176	Na_2O	Na	0.7423	9.8706
Cl	O	0.2257	9.3535	Na_2O	NaCl	1.8840	0.2751
CO_2	Na_2CO_3	2.4110	0.3822	Na_2O	Na_2CO_3	1.7080	0.2326
Fe_2O_3	$FeCO_3$	1.1451	0.1615	Na_2O	Na_2SO_4	2.2900	0.3597
KCl	K_2O	0.6320	9.8007	Na_2SO_4	Na	0.3243	9.5109
KCl	K_2SO_4	1.1690	0.0676	Na_2SO_4	Na_2CO_3	0.7463	9.8729
K_2O	K_2CO_3	1.4670	0.1664	Na_2SO_4	Na_2O	0.4368	9.6403
K_2O	K_2SO_4	1.8490	0.2669	SiO_2	CO_2	0.7286	9.8625
$MgCO_3$	Mg	0.2888	9.4606	SO_3	$CaSO_4$	1.7000	0.2306
MgO	$MgCO_3$	2.0900	0.3202	SO_3	K_2SO_4	2.1780	0.3380
MgO	$MgSO_4$	2.9840	0.4747	SO_3	$MgSO_4$	1.5040	0.1773
Mn_3O_4	$MnCO_3$	1.5070	0.1780	SO_3	Na_2SO_4	1.7760	0.2494

CHAPTER XXXIII.

COAL AND COKE.

1. The proximate analysis of coal and coke comprises determinations for MOISTURE, VOLATILE COMBUSTIBLE MATTER, FIXED CARBON and ASH. SULPHUR and PHOSPHORUS are sometimes required in addition.

The results obtained for moisture, volatile combustible matter and fixed carbon depend to a considerable extent upon the particular plan of working adopted by the operator. While the usual methods are more or less conventional and arbitrary, there is as yet no uniform agreement as to details. The methods given below are those adopted in the report of the committee on coal analysis of the American Chemical Society.*

2. **Preparation of the Sample.**—As soon as the coal or coke is received crush it sufficiently and quarter it down to about 100 grams. Grind this coarsely, about as fine as is possible with an ordinary coffee-mill. Transfer a portion of this sample at once to a tightly stoppered bottle for use in determining moisture

Grind 12 to 15 grams of the remainder moderately fine and transfer to a tightly stoppered bottle for use in determinations other than moisture.

3. **Moisture.** — Dry 1 gram of the coal, uncovered, in a weighed porcelain or platinum crucible at 104°–107° C. for one

* Jour. Am. Chem. Soc., XXI, 1116.

hour, best in a double-walled bath containing pure toluene. Cool in a desiccator and weigh covered. Loss of toluene may be prevented by attaching a reflux condenser to the steam vent of the bath.

With coals high in moisture and in all cases where accuracy is desired, determinations must be made both with the coarsely ground and with the powdered coal. When, as will usually be the case, more moisture is found in the coarsely ground than in the powdered coal, a correction must be applied to all determinations made with the latter. The correction factor may be found as follows: Divide the difference in moisture by the per cent. of other constituents than moisture, as found in the powdered coal. Multiply the per cent. of each constituent as found in the powdered coal by the quotient, and subtract the resulting product from the amount of the given constituent.

Thus, suppose the results of an analysis give:

	Coarsely Ground Coal.	Powdered Coal.
Moisture	12.07	10.39
Volatile combustible matter	....	34.25

then the correction factor will be

$$\frac{12.07-10.39}{100-10.39}=\frac{1.68}{89.61}=0.0187$$

and the true per cent. of volatile combustible matter will be

$$34.25-(34.25\times 0.0187)=33.61.$$

4. **Volatile Combustible Matter.**—Place 1 gram of the fresh, undried, powdered coal in a 20- or 30-gram weighed platinum crucible having a tightly fitting cover. Heat over the full

flame of a Bunsen burner for seven minutes. The crucible should be supported on a platinum or pipe-stem triangle with the bottom 6 to 8 cm. above the top of the burner. The flame should be fully 20 cm. high when burning free, and the determination should be made in a place free from draughts. The upper surface of the cover should burn clear, but the under surface should remain covered with carbon. To find "volatile combustible matter" subtract the per cent. of moisture from the loss found here.

The term "Volatile Combustible Matter" does not represent any definite compound or class of compounds which exist in the coal before heating. The method is arbitrary and the application of higher temperatures will expel more volatile matter.

(*Note.*—I have found western lignites to suffer a serious mechanical loss when subjected at once to the full heat of a Bunsen burner as directed above, the loss sometimes amounting to several per cent. To avoid this, with such coals, I proceed as follows: Heat the covered crucible over a Bunsen burner with the flame turned very low and not touching the crucible. After heating for several minutes in this manner, increase the temperature very gradually and only as fast as the coal will permit without suffering mechanical loss. Finally, when the full power of the Bunsen flame is attained adjust it as directed above and continue the heating for five minutes. Cool in desiccator and weigh as usual.)

5. **Ash.**—Burn the portion of powdered coal used for the determination of moisture, at first over a very low flame, with the crucible open and inclined, till free from carbon. If properly treated this sample can be burned much more quickly than the dense carbon left from the determination of volatile matter. It is advisable to examine the ash for unburned carbon by moistening it with alcohol.

When the sulphur in the coal is in the form of pyrites, that compound is converted almost entirely into ferric oxide in the determination of ash, and, since three atoms of oxygen replace four atoms of sulphur, the weight of the ash is less than the weight of the mineral matter in the coal by five-eighths of the weight of the sulphur. While the error from this source is sometimes considerable, the committee does not recommend such a correction for "proximate" analyses. When analyses are to be used as a basis for calculating the heating effect of the coal a correction should be made.

6. **Fixed Carbon.**—This is found by subtracting the per cent. of ash from the per cent. of coke or residue left after expelling the volatile matter. Or, it is the difference between the sum of the other constituents determined and 100. Sulphur, which passes partly into the "volatile combustible matter" and partly into the coke, is not considered in the calculation.

7. **Coking Quality.** — (Not included in committee's report). The condition of the residue left in the crucible after expelling the volatile matter is a good indication of the coking qualities of a coal, and the latter may usually be reported as "coking" or "non-coking" accordingly. A better test, however, is to place a quantity of the coal in the shape of small fragments in a clay crucible, cover the latter and expose to a strong heat in a muffle. Be careful not to fill the crucible too full or the mass may swell up and displace the cover. When all volatile matter has apparently been expelled and the crucible is at a bright red heat, remove from the muffle and allow to cool with the cover on. When cold, examine the residue, and from its fused or unfused condition, its porosity or density, etc., decide as to the quality of the coke. Of course this test is unnecessary with coals, such as lignites, that give only a loose non-coherent residue in the platinum crucible after expelling the volatile matter.

8. Sulphur.—This is determined by Eschka's method as follows: Weigh 1 gram of the finely powdered coal into a platinum dish of 75- to 100-cc. capacity. Add 1.5 grams of an intimate mixture of 1 part dry sodium carbonate and 2 parts magnesium oxide. The magnesium oxide should be of a light and porous nature, not compact and heavy. Mix the coal and other constituents in the dish very thoroughly with a platinum spatula or glass rod. Now, holding the Bunsen burner in the hand at first, heat the dish very cautiously and raise the temperature very slowly, especially with soft coals. When the strong glowing has ceased, increase the heat gradually, placing the burner under the dish, until in about fifteen minutes the bottom of the dish is at a low red heat.*

Stir the mixture occasionally with a platinum rod and when the carbon is completely burned allow to cool and transfer the residue to a beaker. Rinse out the dish into the beaker with about 50 cc. of water, add 15 cc. of saturated bromine water and boil the mixture for five minutes. Remove from the heat, allow the insoluble matter to settle and decant the solution through a filter. Treat the residue twice with 30 cc. of water, bringing it to a boil each time, allowing to settle and decanting through the filter. Finally, transfer the residue to the filter and wash until the filtrate gives only a slight opalescence with nitric acid and silver nitrate. Add to the filtrate, which should have a volume of about 200 cc., 1.5 cc. of strong hydrochloric acid and boil the solution until the bromine is expelled. Now add to the boiling liquid 10 cc. of a 10 per cent. solution of barium chloride. This should be added drop by drop, especially at first, with constant stirring. Allow to stand on a hot plate or over a low flame until the solution is perfectly clear, and then

* To avoid possible absorption of sulphur from the illuminating gas it is ad visable to place the dish in a hole cut in a piece of asbestos board.

filter off the barium sulphate, washing with hot water until free from chlorides. Transfer the filter and moist precipitate to a weighed platinum crucible and heat with a low flame until the paper is burned. Finally heat to redness, cool in a desiccator and weigh the $BaSO_4$. The weight of the latter, multiplied by 0.1373, will give the weight of the sulphur, from which the percentage may be calculated.

To insure accuracy a blank determination should be made, using all the reagents in the same quantities and carrying out the entire process in exactly the same manner as with the coal. Subtract the weight of any barium sulphate found from that obtained in the coal test and calculate the true percentage of sulphur in the coal from the remainder.

If the coal contains much pyrite or calcium sulphate, some sulphur may still remain with the residue of magnesium oxide and ash. Dissolve this residue in hydrochloric acid in slight excess, dilute sufficiently, heat to boiling and precipitate as above with barium chloride. Filter off any barium sulphate obtained, ignite, and weigh as before, and add the weight to that of the main portion. The experiments of Geo. Steiger, under the direction of Dr. Hillebrand, demonstrate the necessity of always examining this residue; 9 samples of coal showing from 0.032 to 0.114 per cent. additional sulphur, where the total sulphur ran from 0.625 to 4.561 per cent.

9. Error in Determining Volatile Combustible Matter.—Mead and Attix* call attention to the fact that while the above described method of heating over a Bunsen burner may be sufficient for soft coals it is quite inadequate for coke or anthracite which require the heat of a blast-lamp to drive off all the volatile combustible matter. They further show that during this heating

* Jour. Am. Chem. Soc., XXI, 1137.

there is a considerable loss by oxidation. With coke and anthracite they therefore proceed as follows: Heat in an atmosphere of nitrogen, using a Shimer crucible,* first over a Bunsen burner for several minutes and then over a blast for fifteen minutes. As an alternative method, giving nearly the same results, omit the atmosphere of nitrogen but make two similar heatings and weighings, using an ordinary covered platinum crucible. The loss in a series of several consecutive heatings after the first is shown to be practically a constant, and is due entirely to oxidation. Therefore, deduct the loss sustained in the second heating from the loss in the first to obtain the loss due only to volatile combustible matter.

This method is undoubtedly much more satisfactory for coke and anthracite than the one given above, but as the results in any case are intended to be only comparative, it is perhaps not advisable to adopt it for ordinary work until there is some general agreement in the matter.

10. **Phosphorus.**—Weigh 5 grams of the coarsely crushed coal or coke (2) into a platinum dish and burn completely to ash, best in a muffle. Transfer the residue to a beaker and digest with hydrochloric acid for some time to dissolve the phosphorus compounds, then filter into an 8-oz. beaker, washing with water, and evaporate the filtrate to dryness on a sand-bath or hot plate. Add to the residue in the beaker 25 cc. of strong nitric acid, cover the beaker and boil to about 12 cc. Now dilute with 12 cc. of water and filter, washing with water. Receive the filtrate in a 6-oz. flask. The total volume should not exceed 50 cc. Heat to 40°–45° C., add 60 cc. of molybdate solution (see XXII, 7), previously filtered and heated to 40°–45° C., stopper the flask and shake for five minutes. Allow to stand in a warm place for fifteen

* Jour. Am. Chem. Soc., XXI, 557.

minutes and then filter through a 7-cm. washed filter (Baker and Adamson, A grade) that has been previously dried at 110° C. and weighed. Wash with a 2 per cent. nitric acid solution till free from salts, and then twice with 95 per cent. alcohol. Dry twenty minutes at 110° C. and weigh. The dried residue contains 1.63 per cent. of phosphorus, therefore multiply the weight found by 0.0163 to obtain the weight of phosphorus in the amount of coal taken.

Always test the filtrate with more molybdate solution, omitting the washings.

11. **Specific Gravity.**—In the matter of specific gravity either one of two things may be required:

1. The *true* specific gravity, or specific gravity of the particles.
2. The *apparent* specific gravity, or specific gravity of the mass taken as a whole.

Proceed as follows:

Select a lump of convenient size, weighing, perhaps, from 10 to 20 grams, and weigh it. Call this weight *a*.

Place the piece in distilled water under a bell-jar and exhaust the air from the latter. Admit the air, carefully wipe off the surface of the coal, now saturated with water, and weigh again. Call the increase in weight, or weight of the absorbed water, *b*.

Now suspend the wet coal by means of a fine horse-hair and weigh it immersed in water. Call this weight *c*.

Then,

$$\frac{a}{a-c} = \text{true specific gravity,}$$

and

$$\frac{a}{a-(c-b)} = \text{apparent specific gravity.}$$

12. Heating Value of Coal.—The heating value of a coal is usually expressed either in terms of British Thermal Units (abbreviated, B.T.U.), or in Calories (abbreviated to Cal.).

The British Thermal Unit is the heat required to raise the temperature of 1 pound of pure water, at or near its temperature of maximum density, 39.1 F., through 1 degree Fahrenheit.

The Calorie is the heat necessary to raise the temperature of 1 kilogram of water from 4° C. to 5° C.

1 B.T.U. is equivalent to 0.252 Calorie.

1 Calorie is equivalent to 3.968 B. T. U.

To reduce B.T.U. per pound to Calories per kilogram, multiply by 5/9.

To reduce Calories per kilogram to B.T.U. per pound, multiply by 9/5.

The heating value of a coal is most accurately determined by means of a coal calorimeter, and a fairly correct result may be obtained by calculation from an ultimate analysis. Descriptions of both these methods may be found by consulting the standard authorities, but are omitted here as beyond the present scope of this treatise.* Attempts have been made to calculate the heating value of a coal from the results of its proximate analysis, and the following formulas are given as sufficiently accurate for rough heat calculations.

13. Formula of F. Haas (West Virginia Geol. Survey).

B.T.U. per pound = 156.75 (100 − % ash − % sulphur − % moisture) + (40.50 × % sulphur).

Calories per kilogram = 87.1 (100 − % ash − % sulphur − % moisture) + (25.50 × % sulphur).

14. Goutal's Formula.†

$82C + aV$ = Calories per kilogram.

---

* See Chemical Engineer, Vol. I, pp. 1–98, for a clearly written article.

† Jour. f. Gasbeleuchtung u. Wasserorgung, XLVIII, 1006.

C = Fixed Carbon, V = Volatile Constituents, *a* is a variable factor dependent upon the Volatile Constituents, V_1, of the pure (water-and ash-free) coal.

$$V_1 = \frac{V \times 100}{C + V}$$

15. The following table gives the values of *a* corresponding to values of V_1 from 1 to 40.

V_1	*a*	V_1	*a*	V_1	*a*	V_1	*a*	V_1	*a*	V_1	*a*	V_1	*a*
1–5	150	10	130	16	115	21	108	26	102	31	97	36	91
5	145	11	127	17	113	22	107	27	101	32	97	37	88
6	142	12	124	18	112	23	105	28	100	33	96	38	85
7	139	13	122	19	110	24	104	29	99	34	95	39	82
8	136	14	120	20	109	25	103	30	98	35	94	40	80
9	133	15	117										

16. As an example of how the results obtained by the above formulas compare with that of the bomb calorimeter, the following instance is given:

Proximate Analysis of Coal.

Moisture,	1.460
Volatile,	10.880
Fixed Carbon,	81.535
Ash,	6.125
	100.000
Sulphur,	0.807

Calories per kilogram determined by calorimeter, 7778.

By Haas's formula:

$$87.1\ (100 - 6.125 - 0.807 - 1.46) + (22.5 \times 0.807) = 7909$$

Cal. per kilo.

By Goutal's formula:

$$V_1 = \frac{10.88 \times 100}{81.535 + 10.88} = 11.78.$$

From the table we find that with V_1 equal to 11.78, $a = 124.7$
Then,

$$82\ (81.535) + 124.7\ (10.88) = 8068 \text{ Cal. per kilo.}$$

CHAPTER XXXIV.

TESTING CRUDE PETROLEUM.

Owing to frequent discoveries of petroleum in new districts the western technical chemist is liable to be required to test and compare samples with a view of determining their commercial value. The following tests will usually suffice:

1. **Preliminary Note.** — The sample as received is likely to contain water, either from the original flow being mixed with water, or from the containing vessel having been rinsed out with water and not subsequently dried. Therefore, allow the sample to stand for some time and settle out as much as possible before beginning the tests.

2. **Specific Gravity.** — This may be taken with the hydrometer if there is a sufficient quantity of the oil at hand. Fill the hydrometer jar about four-fifths full and introduce the hydrometer carefully so as not to smear the stem with oil above the point to which it would naturally sink. The hydrometer to employ is the "Standard Baumé Hydrometer for Coal Oil." Hold a piece of white paper back of the jar and note the point where the lower portion of the meniscus touches the scale. Take also the temperature of the oil and subtract 1° Baumé from the hydrometer reading for every increase of 10° Fahr., or 5.5° C., above 60° Fahr., or 15.5° C. The specific gravity may be found by

comparing the hydrometer reading with a table (see p. 341), or by the formula $\frac{144.3}{134.3+B^\circ}$. "B°" is the reading Baumé.

If the quantity of oil is too small for filling a hydrometer jar the specific gravity may be taken with a 10- or 25-cc. picnometer. The picnometer should be tested with water at 15.5° C. and then with the oil at the same temperature. The weight of the oil divided by the weight of the same volume of water, both at 15.5° C., will give the specific gravity at this temperature.

3. **Distillation Test, Engler's Method.**—Arrange a fractional distillation-flask with a side tube and Liebig condenser as in the figure. The apparatus should be approximately of the dimensions shown. If possible, have the side tube long enough to permit of bending the outer end downward so as to fit into the vertical condenser. When this cannot be done without bringing the condenser and receiver too near the flame, use a properly bent glass tube as an extension. The joints may be cork, or strips of cotton cloth moistened with glycerine wrapped around and wound tightly with small string. Rubber will not answer so well, as it is acted upon by the hot vapors and may swell up or split. A Bunsen burner or alcohol lamp may be used for heating. A piece of wire gauze may be placed under the flask at first and afterward removed and the naked flame employed as a higher temperature is required. The flame should be screened to protect it from draughts.

4. By means of a graduated pipette, place 100 cc. of the oil in the flask and then arrange a thermometer, graduated to 400° C., inserted through a cork in the top of the flask, so that the top of the bulb is about opposite the junction of the side tube and neck of the flask. Heat the oil so that it distils as nearly as possible at the rate of 2 to 2.5 cc. per minute. As the temperature rises above 100° C. be watchful for water in the oil. Water is apt to cause sudden violent bumps when it collects

under the oil. It also evaporates and a portion condenses in a drop on the thermometer bulb. When this drop falls back into

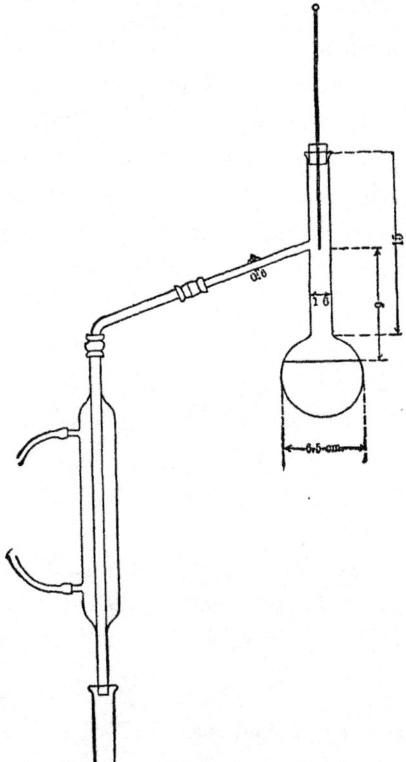

FIG. 21

the hot oil it is liable to cause a small explosion and perhaps spoil the analysis. It is best to heat very cautiously at first, and if water appears on the bulb remove the thermometer and absorb it with filter-paper.* Oils with but little naphtha are apt to give the most trouble in this respect, as they may become very hot

* A safer plan is to lower the thermometer so the bulb is immersed in the liquid, until a temperature of, say, 120° C. is attained, occasionally agitating the flask until all water is expelled. Then raise the thermometer to its proper place and proceed as described.

and dangerous before the thermometer in the neck shows much rise, and if a drop of water then falls back a violent burst of steam is the result. Continue the heating until the thermometer indicates a temperature of 150° C. Collect the distillate in a 50-cc. cylindrical graduate. When a temperature of 150° is reached, remove the lamp a moment so that the temperature will drop back 20°, then replace it and again heat to 150°. Repeat these operations until no appreciable additional distillate is thus obtained. Read the number of cubic centimeters in the graduate, quickly pour off the oil and replace the graduate again. Now increase the heat so that the thermometer gradually rises to 200° and then repeat the cooling and heating as before. Measure the fraction thus obtained and then increase the heat similarly to 250° and 300°,* and then above 300°, until nothing but coke remains in the flask. The distillate up to 150° is called naphtha. The next three portions are illuminating oils, the fractions obtained showing the proportions of the different grades. The distillate coming off above 300° is called lubricating oil, and what is finally left in the flask constitutes the residuum and coke. The distillation is usually carried as far as possible.

EXAMPLE SHOWING TABULATION OF RESULTS.

Specific gravity of the crude oil at 60° Fahr., 0 841.

Temperature of Distillation.	Per Cent. of Product.		Nature of Product.
Below 150° C.	28.5		Naphtha.
From 150° to 200°	12.5	Total 66%.	Illuminating-oils.
" 200° " 250°	13.0		
" 250° " 300°	41.0		
Above 300°	2.0		Lubricating-oil.
Residue.	3.0		Coke.
	100.0		

* I have found it best, at this point, to stop the flow of cold water and cautiously heat that standing in the condenser, with a soft flame, so as to prevent subsequent products from thickening or solidifying in the tube.

MISCELLANEOUS.

Determination of Antimony and Tin in Babbitt, Type Metal, and Other Alloys.*—The following process, found generally applicable to alloys and to the sulphides of antimony and tin, either in the solid state or in proper solution, claims nothing in particular as original, except the direct determination of the antimony and tin in one portion of the alloy without separating the other ingredients, and the titration of antimony and tin when both are present in solution. The direct titration in an alloy without separation in the other ingredients may not be applicable in all cases, but it is in most cases, and where it is not the sulphide separation mentioned below is made use of and the sulphides treated in almost the same way as the original alloy. If other metals, besides tin and antimony, are to be determined, Rossing's method given below gives a good, quick separation, but the antimony and tin are determined by direct titration, if possible, in another portion of the alloy.

Where only antimony and tin are sought, the alloy may be decomposed by nitric acid, by sulphuric acid, or by a mixture of sulphuric acid and potassium sulphate. Where possible, sulphuric acid alone is used. If nitric acid is first used, it must subsequently be expelled by boiling with sulphuric acid, and after the nitric acid is expelled, some tartaric acid and some potassium sulphate must be added, and the melt heated till all carbon has been oxidized. This leaves the antimony and tin in the proper state for the titrations.

The two standard solutions required are an N/10 potassium permanganate, and an N/10 iodine solution. For antimony determination, the permanganate solution should be standardized by C.P. antimony or tartar emetic (anhydrous) of the proper

* Method of W. H. Low, Jour. Am. Chem. Soc., XXIX, 66.

composition. If properly done, these standards will agree and both will agree with a standardization made with oxalic acid in the regular way, but always standardize with antimony or its compounds by heating with sulphuric acid, diluting and adding the same amount of hydrochloric acid as is used with the alloy. In this way correct results are insured. The iodine solution is compared with a solution of sodium thiosulphate of known value.

From 0.5 gram to 1 gram of the finely divided alloy is taken, and as the method is generally the same, the standardization of the permanganate will be described together with the subsequent titration of the tin.

0.1202 gram of finely powdered C.P. antimony (= 0.3234 gram tartar emetic) and 0.1190 gram of tin are placed in a 450 cc. Jena glass Erlenmeyer flask and about 10 cc. of strong sulphuric acid (free from chlorine compounds) added; 3-4 grams of potassium sulphate are sometimes used and can be added here. Heat till the metals are all in solution (or the alloy thoroughly decomposed) and all separated sulphur has been boiled out of the flask. All sulphurous acid will also be expelled by this time. Do not drive off all free sulphuric acid, but have enough left to keep the melt from getting hard on cooling. About 7-10 cc. left is enough. These operations take only a few minutes. Cool and add 50 cc. of water and 10 cc. of strong hydrochloric acid and heat to get all possible in solution. Large quantities of lead sulphate, even, will dissolve, and the solution will become clear. However, the object is to get the antimony and tin in solution, and this is all that is necessary. With the quantity of antimony taken no tartaric acid is necessary, but with larger quantities a few grams of tartaric acid must be added. Tartaric acid has no effect on either titration. Some stannic compound with sulphuric acid may go into solution with some difficulty, but no attention need be paid to it here, as it will dissolve when we get through with the antimony, if not before. Cool the solution and add about 110 cc. more of water and 25–30 cc. (with small amounts of Sb, such as is in solder, a total of 20 cc. strong HCl in total volume of 200 cc. appears enough) more of strong hydrochloric acid. Thoroughly cool this mixture and proceed to titrate with permanganate. Add the latter till the last drop colors the whole solution pink. The end point is sharp, but the color may not remain long, owing to the large quantity of hydrochloric acid present. If less hydrochloric acid were present, say 10 cc. in the total volume of 200 cc., the end point would be sharp, but would apparently occur at about 19.60–19.70 cc., instead of 20.00 cc., as it should. The true end point under these circumstances is troublesome to find. But even an incorrect end point may give good results, if the solution has been standardized in exactly the same way. The determination of the antimony or standardization of the permanganate is now finished and the tin is the next consideration. The titrated solution contains antimonic chloride and stannic chloride, besides other things of no particular moment. Pour this solution into a 500 cc. round-bottomed flask and rinse out the Erlenmeyer flask with about 50 cc. of strong hydrochloric acid and add washings to main solution. The main solution should be at least one-fifth by volume of strong hydrochloric acid (the regular strong C.P. acid). Add about 1 gram of finely powdered C.P. antimony * and place on the steam bath for about 15 minutes, shaking once in a while. Next remove from the bath and connect with an apparatus, capable of giving a rapid current of carbon dioxide. The connection is made by means of a cork carrying two tubes. The first dips below the surface of the solution in the flask and the second or outlet tube is bent downward and the end dips slightly

* Cf. Ibbotson and Brearley, Analyst, 1902, 25.

below the surface of some water or mercury. This allows any tendency to back pressure to be instantly detected. While passing a rapid stream of gas, heat the contents of the flask to boiling, using a naked flame, but taking care to heat the sides of the flask and avoid directly heating the bottom. This is because the antimony lies on the bottom and the flask might be cracked by unequal heating in contact with solution. Boil 2–3 minutes after the liquid commences to boil. Cool quickly by surrounding the flask with cold water and take care that the current of carbon dioxide is strong enough to prevent back pressure due to sudden condensation. When cold loosen the cork somewhat and introduce 5 cc. of good starch solution, and then withdraw the flask gently so as not to allow air to enter by forming currents. Cork quickly and take to the burette. Introduce the spit of the burette far down into the neck of the flask and rotating the latter gently run in or drop in N/10 iodine solution. Towards the end the starch blue will appear and remain mostly in the middle portion of the solution, requiring stronger agitation to mix with the rest of the solution. This causes the metallic antimony in the bottom of the flask to become stirred up and slightly obscure any slight blue tint. For this reason the titration is continued till the last drop gives a strong blue to the whole solution, and then we deduct about 0.05 cc. from the burette reading. Do not fear that the end point will not be recognized within one drop or less. It is unmistakable with good starch solution, and no doubtful ending should be taken. With correct solutions, etc., the titration should have taken just 20.00 cc. Mixtures of C.P. antimony and C.P. tin (allowing for 0.10 per cent. impurities found to be present) give exact results. No trouble has been found in titrating tin correctly with iodine. Objections to this method may be founded on accepting C.P. tin as actually nothing but tin, while most of it contains impurity.

To test tin or antimony for impurity, a quick method of considerable accuracy is to take about 0.5 gram in a porcelain boat, place in a combustion tube and pass a slow current of dry chlorine and dry hydrochloric acid gas. The tin or antimony is quickly volatilized and lead, copper, iron, etc., remain mostly in the boat. Displacing the chlorine with carbon dioxide and then heating in a current of hydrogen reduces the chlorides left in the boat to metal, and their weight can be deducted from the original metal, or the percentage determined. This is not strictly accurate, as on first heating in hydrogen there is a slight volatilization of some of the chlorides.

In the case of an alloy a good qualitative analysis should always be made, unless the approximate composition is known. If there are no interfering metals, the alloy is decomposed and the antimony and tin titrated, as shown above, without removing the lead sulphate, copper sulphate, etc. Lead in large amount does not interfere. Theoretically, any amount of copper should not interfere, and practically small amounts are known to cause no trouble, while large amounts have not been present in the tests made. Should the alloy contain iron in any quantity, there might be some danger of ferrous sulphate being left after boiling with sulphuric acid. This difficulty is overcome by decomposing the alloy with nitric acid, as usual, boiling off most of the nitric acid, adding about 10 cc. of sulphuric acid and 3-4 grams potassium sulphate, boiling to complete expulsion of the nitric acid, and then adding a little tartaric acid or other organic matter, and after the carbonization, continuing the boiling till all organic matter has been destroyed. This will always leave the antimony and tin in the "ous" and "ic" states, respectively.* From this point, the determination is carried on in the manner described

* A. H. Low, Jour. Am. Chem. Soc., XXVIII, 1715.

above. Where the antimony and tin have been separated from the other metals by means of alkaline sulphides, the details of operation depend on whether the decomposition of the alloy was made with aqua regia or with sulphuric or nitric acid. It will be assumed that the alloy has been dissolved in aqua regia and the tin and antimony are in solution combined with alkaline sulphides. Add slight excess of sulphuric acid and heat gently to precipitate the sulphides of antimony and tin. Filter, preferably through a 60 mm. Witt's plate and an S. & S. white ribbon filter No. 589, and wash the precipitate on the filter with dilute hydrogen sulphide water containing enough ammonium acetate to prevent stannic sulphide from giving a turbid or opalescent filtrate. After washing somewhat, test each succeeding 10–20 cc. of filtrate by boiling off the hydrogen sulphide, acidifying with nitric acid and adding three or four drops of silver nitrate solution. The last filtrate should be entirely free from chlorine, as shown by this test. With a Witt's plate of 60 mm. diameter, this is quickly accomplished. The residue on the filter and filter itself may now be placed in a 450 cc. Jena glass Erlenmeyer flask, 3–4 grams potassium sulphate added and 10–20 cc. strong sulphuric acid, and the whole mass boiled till all organic matter has been destroyed and the antimony and tin are present as sulphates, while there are not over 10 cc. of free sulphuric acid left in the flask. As organic matter is not really necessary in this operation, and it takes some time to oxidize the filter paper, the precipitate on the filter can be handled differently, if some alkaline sulphide, free from chlorine compounds, is available. This desideratum will probably be found best in ammonium sulphides or poly-sulphides, as the best C.P. by alcohol caustic soda contains enough chloride to cause some loss of tin. The precipitate is washed from the filter into the flask with as small a quantity of water as possible. What remains on the filter is dissolved off with alkaline sulphide free from chlorine and the solution added to the flask. Enough alkaline sulphide is now added to the flask to dissolve the sulphides of antimony and tin on warming. When this is accomplished, acidify with sulphuric acid and then add about 10–15 cc. in excess and 3–4 grams potassium sulphate and boil down to sulphuric acid fumes and till all sulphur has been expelled and the antimony and tin remain only as sulphates. The final amount of free sulphuric acid left should not be over 10 cc. Tin tends to form a stannic sulphate, very insoluble in strong sulphuric acid, but subsequent heating and boiling in the presence of hydrochloric acid dissolves it. The boiling with strong sulphuric acid and destruction of organic matter is carried on easily and quickly over a naked flame of good heating power. The Jena glass Erlenmeyer flasks stand the operation very well, and none have cracked. The flask is generally held in the hand by means of a clamp (Chaddocks' clamp with rubber removed is very good) and rotated to insure even heating. In this way the sulphuric acid may be completely driven off, leaving a melt of acid potassium sulphate, and the flask will not break. If there is any doubt that the antimony does not exist in the "ous" condition, a little tartaric acid or other organic matter may be added and burned off by boiling. The antimony and tin now exist as sulphate and the procedure is the same as in the standardization of the permanganate.

This method for the sulphides of antimony and tin is in every way much more satisfactory than the gravimetric separation and determination, and takes little time.

Mixtures of C.P. antimony and tin (allowing for any impurities) give exact results. An alloy run through recently gave the following results. Qualitative analysis showed the metals indicated to be the only ones present, except in minute traces.

Lead	74.19	Alloy dissolved in aqua regia, alkaline sulphide separation, lead separated from copper as sulphate in presence of alcohol. Two determinations checked.
Antimony	15.44	Alloy boiled with sulphuric acid and antimony titrated direct, as in standardizing permanganate.
Tin	9.88	Determined in same solution, after the antimony.
Copper	0.44	Determined after the lead, driving off alcohol and determining by A. H. Low's iodide process. Check test.
Iron	trace	
	99.95	

If arsenic were present, it would interfere with the antimony titration or be counted as antimony. It may easily be removed.

It is often stated that the titration of tin with iodine gives slightly low results. Experience with the above method has given confidence that the results for tin are exact, if the conditions are observed. Tin may be lost somewhere in the process, but all that remains is surely indicated by the titration.

Method of A. Rossing for Separation of Sulphides. * — In this method the alloy is dissolved in a minimum of aqua regia, using a little potassium chlorate to insure complete oxidation. After some dilution a little tartaric acid is added, the solution almost exactly neutralized with sodium hydroxide solution and a sufficient quantity of colorless sodium hydrosulphide, NaSH, added to precipitate copper, lead, etc., and to retain all the antimony and tin in solution. By using a gentle heat and with enough sodium sulphide the antimony and tin will go into solution. Filter, washing the residue with hot dilute sodium sulphide solution. Test the final washings by acidifying them, to make sure of the removal of all the antimony and tin from the residue.

* Jour. Soc. Chem. Ind., 1902, 191.

TABLES.

RELATION OF BAUMÉ DEGREES TO SPECIFIC GRAVITY, FOR LIQUIDS LIGHTER THAN WATER.

Baumé.	Specific Gravity.	Baumé.	Specific Gravity.	Baumé.	Specific Gravity.	Baumé.	Specific Gravity.
10	1.0000	31	0.8695	52	0.7692	73	0.6896
11	0.9929	32	0.8641	53	0.7650	74	0.6863
12	0.9859	33	0.8588	54	0.7608	75	0.6829
13	0.9790	34	0.8536	55	0.7567	76	0.6796
14	0.9722	35	0.8484	56	0.7526	77	0.6763
15	0.9655	36	0.8433	57	0.7486	78	0.6730
16	0.9589	37	0.8383	58	0.7446	79	0.6698
17	0.9523	38	0.8333	59	0.7407	80	0.6666
18	0.9459	39	0.8284	60	0.7368	81	0.6635
19	0.9395	40	0.8235	61	0.7329	82	0.6604
20	0.9333	41	0.8187	62	0.7290	83	0.6573
21	0.9271	42	0.8139	63	0.7253	84	0.6542
22	0.9210	43	0.8092	64	0.7216	85	0.6511
23	0.9150	44	0.8045	65	0.7179	86	0.6481
24	0.9090	45	0.8000	66	0.7142	87	0.6451
25	0.9032	46	0.7954	67	0.7106	88	0.6422
26	0.8974	47	0.7909	68	0.7070	89	0.6392
27	0.8917	48	0.7865	69	0.7035	90	0.6363
28	0.8860	49	0.7821	70	0.7000	95	0.6222
29	0.8805	50	0.7777	71	0.6965		
30	0.8750	51	0.7734	72	0.6930		

TABLE OF MEASURES AND WEIGHTS.

Measures of Capacity.

A. Dry Measure.

1 bushel = 2150.42 cubic inches.
1 " = the volume of 77.627 pounds of distilled water at 4° C.
Legal: 1 liter = 0.908 quart.
1 bushel = 4 pecks = 8 gallons = 32 quarts = 35.24229 liters.
1 peck = 2 gallons = 8 quarts = 8.81057 liters.
1 gallon = 4 quarts = 4.40528 liters.
1 quart = 1.10132 liters.

B. Liquid Measure.

1 U. S. gallon = 231 cubic inches.
1 gallon = the volume of 8.3388822 pounds = 58378 troy grains of distilled water at 4° C. (Stillman, Engineering Chemistry.)
1 " = 58318 grains of water at 62° F. (U. S. Phar.)
1 " = 58334.9+ grains of pure water at 60° F., weighed in air at 60° F., at barometric pressure of 30 inches of mercury. (Mason, Examination of Water.)
Legal: 1 liter = 1.0567 quart = 0.26417 gallon.
1 gallon = 4 quarts = 8 pints = 32 gills = 3.78544 liters.
1 quart = 2 pints = 8 gills = 0.94636 liter.
1 pint = 4 gills = 0.47318 liter.
1 gill = 0.118295 liter.
1 cubic foot = 7.48 gallons = 28.315 liters = 62.42 pounds of water at 60° F. (Stillman.)
1 cubic foot of water at 62° F. = 62.355 pounds avoirdupois = 28320 grams.
1 cubic inch of water at 62° F. = 0.0361 pounds avoirdupois = 16.387 grams. (Watts' Dictionary, V, 1010.)

Linear Measure.

1 yard = 0.91440 meter.
1 foot = 0.30480 meter.
1 inch = 0 0254 meter.
39.37 inches = 1 meter.

Weights.

1 grain troy = 0.0648004 gram.
1 pound troy = 0.822857 pounds avoirdupois.
1 pound avoirdupois = 7000 grains troy = 1.215279 pounds troy.

Troy Weight.

1 pound = 12 oz. = 240 pwts. = 5760 grains = 373.2418 grams.
1 oz. = 20 pwts. = 480 grains = 31.1035 grams.
1 pwt. = 24 grains = 1.5552 grams.
1 grain = 0.0648 grams.
1 gram = 15.432 troy grains.

Avoirdupois Weight.

1 ton = 20 hundredweight = 2240 pounds = 1016.04 kilograms.
1 hundredweight = 112 pounds = 50.80 kilograms.
1 pound = 16 ounces = 256 drams = 7000.00 grains = 453.5900 grams.
1 ounce = 16 drams = 437.50 grains = 28.3495 grams.
1 dram = 27.34 grains = 1.7718 grams.
1 net ton = 2000 pounds = 29166$\frac{2}{3}$ ozs. troy = 907.19 kilograms.

Metric Ton.

1 metric ton = 1000 kilograms.

CONVERSION OF THERMOMETER READINGS.

To convert Fahrenheit to Centigrade, subtract 32 and multiply by $\frac{5}{9}$.
To convert Centigrade to Fahrenheit, multiply by $\frac{9}{5}$ and add 32.

INTERNATIONAL ATOMIC WEIGHTS, 1908.*

Aluminum	Al	27.1
Antimony	Sb	120.2
Argon	A	39.9
Arsenic	As	75.0
Barium	Ba	137.4
Bismuth	Bi	208.0
Boron	B	11.0
Bromine	Br	79.96
Cadmium	Cd	112.4
Caesium	Cs	132.9
Calcium	Ca	40.1
Carbon	C	12.0
Cerium	Ce	140.25
Chlorine	Cl	35.45
Chromium	Cr	52.1
Cobalt	Co	59.0
Columbium	Cb	94.0
Copper	Cu	63.6
Erbium	Er	166.0
Europium	Eu	152.0
Fluorine	F	19.0
Gadolinium	Gd	156.0
Gallium	Ga	70.0
Germanium	Ge	72.5
Glucinum	Gl	9.1
Gold	Au	197.2
Helium	He	4.0
Hydrogen	H	1.008
Indium	In	115.0
Iodine	I	126.97
Iridium	Ir	193.0
Iron	Fe	55.9
Krypton	Kr	81.8
Lanthanum	La	138.9
Lead	Pb	206.9
Lithium	Li	7.03
Magnesium	Mg	24.36
Manganese	Mn	55.0
Mercury	Hg	200.0
Molybdenum	Mo	96.0
Neodymium	Nd	143.6
Neon	Ne	20.0
Nickel	Ni	58.7
Nitrogen	N	14.01
Osmium	Os	191.0
Oxygen	O	16.0
Palladium	Pd	106.5
Phosphorus	P	31.0
Platinum	Pt	194.8
Potassium	K	39.15
Praseodymium	Pr	140.5
Radium	Rd	225.0
Rhodium	Rh	103.0
Rubidium	Rb	85.5
Ruthenium	Ru	101.7
Samarium	Sa	150.3
Scandium	Sc	44.1
Selenium	Se	79.2
Silicon	Si	28.4
Silver	Ag	107.93
Sodium	Na	23.05
Strontium	Sr	87.6
Sulphur	S	32.06
Tantalum	Ta	181.0
Tellurium	Te	127.6
Terbium	Tb	159.2
Thallium	Tl	204.1
Thorium	Th	232.5
Thulium	Tm	171.0
Tin	Sn	119.0
Titanium	Ti	48.1
Tungsten	W	184.0
Uranium	U	238.5
Vanadium	V	51.2
Xenon	Xe	128.0
Ytterbium	Yb	173.0
Yttrium	Yt	89.0
Zinc	Zn	65.4
Zirconium	Zr	90.6

* These are the atomic weights used throughout this book. For technical work it was not deemed necessary, at present, to adopt the latest figures, since the changes they would produce in results are very slight.

INTERNATIONAL ATOMIC WEIGHTS, 1918.

Aluminium	Al	27.1
Antimony	Sb	120.2
Argon	A	39.88
Arsenic	As	74.96
Barium	Ba	137.37
Bismuth	Bi	208.0
Boron	B	11.0
Bromine	Br	79 92
Cadmium	Cd	112.40
Caesium	Cs	132.81
Calcium	Ca	40 07
Carbon	C	12.05
Cerium	Ce	140.25
Chlorine	Cl	35.46
Chromium	Cr	52.0
Cobalt	Co	58 97
Columbium	Cb	93 1
Copper	Cu	63.57
Dysprosium	Dy	162 5
Erbium	Er	167 7
Europium	Eu	152 0
Fluorine	F	19 0
Gadolinium	Gd	157 3
Gallium	Ga	69.9
Germanium	Ge	72 5
Glucinum*	Gl	9 1
Gold	Au	197 2
Helium	He	4 0
Holmium	Ho	163 5
Hydrogen	H	1 008
Indium	In	114 8
Iodine	I	126.92
Iridium	Ir	193 1
Iron	Fe	55 84
Krypton	Kr	82.92
Lanthanum	La	139.0
Lead	Pb	207.20
Lithium	Li	6.94
Lutecium	Lu	175 0
Magnesium	Mg	24 32
Manganese	Mn	54.93
Mercury	Hg	200.6

Molybdenum	Mo	96.0
Neodymium	Nd	144.3
Neon	Ne	20.2
Nickel	Ni	58.68
Niton (radium emanation)	Nt	222.4
Nitrogen	N	14.01
Osmium	Os	190.9
Oxygen	O	16.00
Palladium	Pd	106.7
Phosphorus	P	31.04
Platinum	Pt	195.2
Potassium	K	39.10
Praseodymium	Pr	140.9
Radium	Ra	226 0
Rhodium	Rh	102.9
Rubidium	Rb	85.45
Ruthenium	Ru	101.7
Samarium	Sa	150.4
Scandium	Sc	44.1
Selenium	Se	79.2
Silicon	Si	28 3
Silver	Ag	107.88
Sodium	Na	23.00
Strontium	Sr	87.63
Sulfur	S	32.06
Tantalum	Ta	181.5
Tellurium	Te	127.5
Terbium	Tb	159.2
Thallium	Tl	204.0
Thorium	Th	232 4
Thulium	Tm	168.5
Tin	Sn	118.7
Titanium	Ti	48.1
Tungsten	W	184.0
Uranium	U	238.2
Vanadium	V	51.0
Xenon	Xe	130.2
Ytterbium (Neoytterbium)	Yb	173.5
Yttrium	Yt	88.7
Zinc	Zn	65.3
Zirconium	Zr	90 6

* Also called *Beryllium.*

CHEMICAL FACTORS AND THEIR LOGARITHMS.

	Weighed.	Required.	Factor.	Log.*
Aluminum	Al_2O_3	Al	0.5303	9.7245
	$AlPO_4$	Al_2O_3	0.4185	9.6217
Antimony	Sb_2O_4	Sb	0.7897	9.8975
Arsenic	As_2S_3	As	0.6093	9.7848
	“	As_2O_3	0.8043	9.9054
	$Mg_2As_2O_7$	As	0.4827	9.6837
	“	As_2O_3	0.6372	9.8043
	“	As_2O_5	0.7402	9.8694
Barium	$BaSO_4$	Ba	0.5885	9.7698
	“	BaO	0.6570	9.8176
	$BaCO_3$	Ba	0.6960	9.8426
	“	BaO	0.7771	9.8905
	$BaCrO_4$	Ba	0.5420	9.7340
	“	BaO	0.6050	9.7818
Bismuth	Bi_2O_3	Bi	0.8966	9.9526
	BiOCl	Bi	0.8017	9.9040
	“	Bi_2O_3	0.8941	9.9513
Bromine	AgBr	Br	0.4256	9.6290
	“	HBr	0.4309	9.6344
Cadmium	CdO	Cd	0.8754	9.9422
	CdS	Cd	0.7780	9.8910
	“	CdO	0.8888	9.9488
	$CdSO_4$	Cd	0.5391	9.7317
Calcium	CaO	Ca	0.7148	9.8542
	$CaCO_3$	Ca	0.4006	9.6027
	“	CaO	0.5604	9.7485
	$CaSO_4$	Ca	0.2945	9.4691
	“	CaO	0.4120	9.6149
Carbon	CO_2	C	0.2727	9.4357
Chlorine	AgCl	Cl	0.2472	9.3931
	“	HCl	0.2543	9.4053
	Ag	Cl	0.3285	9.5165
	“	HCl	0.3378	9.5287
Chromium	Cr_2O_3	Cr	0.6846	9.8355
	“	CrO_3	1.3154	0.1190
	“	CrO_4	1.5257	0.1834
	$PbCrO_4$	Cr	0.1613	9.2076
	“	CrO_3	0.3099	9.4913
	“	CrO_4	0.3594	9.5556
	$BaCrO_4$	Cr	0.2055	9.3129
	“	CrO_3	0.3949	9.5964
	“	CrO_4	0.4580	9.6609
Cobalt	Co	CoO	1.2712	0.1042
	$CoSO_4$	Co	0.3805	9.5804
	“	CoO	0.4838	9.6846
Copper	Cu	CuO	1.2516	0.0974
	CuO	Cu	0.7990	9.9025
	Cu_2S	Cu	0.7987	9.9024
	“	CuO	0.9996	9.9999
	$Cu_2(CNS)_2$	Cu	0.5229	9.7184

* The −10 after the logarithms is omitted.

CHEMICAL FACTORS AND THEIR LOGARITHMS—*Continued.*

	Weighed.	Required.	Factor.	Log.
Fluorine	CaF_2	F	0.4866	9.6871
Hydrogen	H_2O	H	0.1119	9.0488
Iodine	AgI	I	0.5405	9.7328
	"	HI	0.5448	9.7362
Iron	Fe_2O_3	Fe	0.6996	9.8449
	"	FeO	0.8999	9.9542
Lead	PbO	Pb	0.9282	9.9676
	PbO_2	Pb	0.8660	9.9375
	PbS	Pb	0.8658	9.9374
	"	PbO	0.9328	9.9698
	$PbSO_4$	Pb	0.6829	9.8344
	"	PbO	0.7357	9.8667
	$PbCl_2$	Pb	0.7447	9.8720
	"	PbO	0.8024	9.9044
	$PbCrO_4$	Pb	0.6406	9.8066
	"	PbO	0.6901	9.8389
Magnesium	MgO	Mg	0.6036	9.7807
	$Mg_2P_2O_7$	Mg	0.2187	9.3399
	"	MgO	0.3624	9.5592
	$MgSO_4$	Mg	0.2023	9.3060
	"	MgO	0.3352	9.5252
Manganese	Mn_3O_4	Mn	0.7205	9.8576
	"	MnO	0.9301	9.9685
	MnS	Mn	0.6317	9.8005
	"	MnO	0.8155	9.9114
	$Mn_2P_2O_7$	Mn	0.3873	9.5880
	"	MnO	0.5000	9.6990
	$MnSO_4$	Mn	0.3641	9.5612
	"	MnO	0.4700	9.6721
Mercury	Hg	HgO	1.0800	0.0334
	HgS	Hg	0.8618	9.9354
	"	HgO	0.9308	9.9689
	HgCl	Hg	0.8494	9.9291
	"	HgO	0.9174	9.9626
Molybdenum	MoO_3	Mo	0.6667	9.8239
Nickel	Ni	NiO	1.2726	0.1047
	NiO	Ni	0.7858	9.8953
	$NiSO_4$	Ni	0.3793	9.5790
	"	NiO	0.4827	9.6837
Nitrogen	$(NH_4)_2PtCl_6$	N	0.0630	8.7993
	"	NH_3	0.0767	8.8848
	"	NH_4	0.0812	8.9096
	Pt	N	0.1438	9.1577
	"	NH_3	0.1749	9.2427
	"	NH_4	0.1852	9.2677
Phosphorus	$Mg_2P_2O_7$	P	0.2784	9.4446
	"	PO_4	0.8531	9.9310
	"	P_2O_5	0.6376	9.8045
	$(NH_4)_3PO_4.12MoO_3$	P	0.0165	8.2178
	"	PO_4	0.0506	8.7042
	"	P_2O_5	0.0378	8.5777

CHEMICAL FACTORS AND THEIR LOGARITHMS—*Continued.*

	Weighed.	Required.	Factor.	Log.
Potassium.	KCl.	K	0 5248	9 7200
	"	K_2O	0 6320	9 8007
	K_2PtCl_6.	K	0.1612	9.2073
	"	K_2O	0.1941	9 2880
	K_2SO_4.	K	0 4491	9.6523
	"	K_2O	0.5408	9.7331
Selenium.	Se.	SeO_2	1.4040	0 1474
Silicon.	SiO_2	Si	0 4702	9.6723
Silver.	AgCl.	Ag	0.7528	9.8766
	AgBr.	Ag	0.5744	9.7592
	AgI.	Ag	0 4597	9.6625
	Ag_2S.	Ag	0.8707	9 9399
Sodium.	NaCl.	Na	0.3940	9.5955
	"	Na_2O	0.5308	9.7249
	Na_2SO_4.	Na	0.3243	9.5109
	"	Na_2O	0.4368	9 6403
	Na_2CO_3.	Na	0.4345	9.6380
	"	Na_2O	0.5853	9.7674
Strontium.	$SrCO_3$.	Sr	0.5935	9.7734
	"	SrO	0.5641	9.8463
	$SrSO_4$.	Sr	0.4770	9.6785
	"	SrO	0.5513	9.7513
Sulphur.	$BaSO_4$.	S	0.1373	9.1377
	"	SO_2	0.2744	9 4384
	"	SO_3	0.3429	9 5352
	"	SO_4	0.4115	9.6143
	"	H_2SO_4	0.4201	9.6233
Tin.	SnO_2.	Sn	0.7880	9 8965
Titanium.	TiO_2.	Ti	0.6005	9.7785
Uranium.	$(UO_2)_2P_2O_7$.	U	0.6671	9.8242
	"	U_3O_8	0.7877	9 8964
	U_3O_8.	U	0.8482	9 9285
Zinc.	ZnO.	Zn	0.8035	9 9050
	ZnS.	Zn	0.6710	9.8267
	"	ZnO	0.8352	9.9218

LOGARITHMS.

Natural Numbers	0	1	2	3	4	5	6	7	8	9	Proportional Parts. 1	2	3	4	5	6	7	8	9
10	0000	0043	0086	0128	0170	0212	0253	0294	0334	0374	4	8	12	17	21	25	29	33	37
11	0414	0453	0492	0531	0569	0607	0645	0682	0719	0755	4	8	11	15	19	23	26	30	34
12	0792	0828	0864	0899	0934	0969	1004	1038	1072	1106	3	7	10	14	17	21	24	28	31
13	1139	1173	1206	1239	1271	1303	1335	1367	1399	1430	3	6	10	13	16	19	23	26	29
14	1461	1492	1523	1553	1584	1614	1644	1673	1703	1732	3	6	9	12	15	18	21	24	27
15	1761	1790	1818	1847	1875	1903	1931	1959	1987	2014	3	6	8	11	14	17	20	22	25
16	2041	2068	2095	2122	2148	2175	2201	2227	2253	2279	3	5	8	11	13	16	18	21	24
17	2304	2330	2355	2380	2405	2430	2455	2480	2504	2529	2	5	7	10	12	15	17	20	22
18	2553	2577	2601	2625	2648	2672	2695	2718	2742	2765	2	5	7	9	12	14	16	19	21
19	2788	2810	2833	2856	2878	2900	2923	2945	2967	2989	2	4	7	9	11	13	16	18	20
20	3010	3032	3054	3075	3096	3118	3139	3160	3181	3201	2	4	6	8	11	13	15	17	19
21	3222	3243	3263	3284	3304	3324	3345	3365	3385	3404	2	4	6	8	10	12	14	16	18
22	3424	3444	3464	3483	3502	3522	3541	3560	3579	3598	2	4	6	8	10	12	14	15	17
23	3617	3636	3655	3674	3692	3711	3729	3747	3766	3784	2	4	6	7	9	11	13	15	17
24	3802	3820	3838	3856	3874	3892	3909	3927	3945	3962	2	4	5	7	9	11	12	14	16
25	3979	3997	4014	4031	4048	4065	4082	4099	4116	4133	2	3	5	7	9	10	12	14	15
26	4150	4166	4183	4200	4216	4232	4249	4265	4281	4298	2	3	5	7	8	10	11	13	15
27	4314	4330	4346	4362	4378	4393	4409	4425	4440	4456	2	3	5	6	8	9	11	13	14
28	4472	4487	4502	4518	4533	4548	4564	4579	4594	4609	2	3	5	6	8	9	11	12	14
29	4624	4639	4654	4669	4683	4698	4713	4728	4742	4757	1	3	4	6	7	9	10	12	13
30	4771	4786	4800	4814	4829	4843	4857	4871	4886	4900	1	3	4	6	7	9	10	11	13
31	4914	4928	4942	4955	4969	4983	4997	5011	5024	5038	1	3	4	6	7	8	10	11	12
32	5051	5065	5079	5092	5105	5119	5132	5145	5159	5172	1	3	4	5	7	8	9	11	12
33	5185	5198	5211	5224	5237	5250	5263	5276	5289	5302	1	3	4	5	6	8	9	10	12
34	5315	5328	5340	5353	5366	5378	5391	5403	5416	5428	1	3	4	5	6	8	9	10	11
35	5441	5453	5465	5478	5490	5502	5514	5527	5539	5551	1	2	4	5	6	7	9	10	11
36	5563	5575	5587	5599	5611	5623	5635	5647	5658	5670	1	2	4	5	6	7	8	10	11
37	5682	5694	5705	5717	5729	5740	5752	5763	5775	5786	1	2	3	5	6	7	8	9	10
38	5798	5809	5821	5832	5843	5855	5866	5877	5888	5899	1	2	3	5	6	7	8	9	10
39	5911	5922	5933	5944	5955	5966	5977	5988	5999	6010	1	2	3	4	5	7	8	9	10
40	6021	6031	6042	6053	6064	6075	6085	6096	6107	6117	1	2	3	4	5	6	8	9	10
41	6128	6138	6149	6160	6170	6180	6191	6201	6212	6222	1	2	3	4	5	6	7	8	9
42	6232	6243	6253	6263	6274	6284	6294	6304	6314	6325	1	2	3	4	5	6	7	8	9
43	6335	6345	6355	6365	6375	6385	6395	6405	6415	6425	1	2	3	4	5	6	7	8	9
44	6435	6444	6454	6464	6474	6484	6493	6503	6513	6522	1	2	3	4	5	6	7	8	9
45	6532	6542	6551	6561	6571	6580	6590	6599	6609	6618	1	2	3	4	5	6	7	8	9
46	6628	6637	6646	6656	6665	6675	6684	6693	6702	6712	1	2	3	4	5	6	7	7	8
47	6721	6730	6739	6749	6758	6767	6776	6785	6794	6803	1	2	3	4	5	5	6	7	8
48	6812	6821	6830	6839	6848	6857	6866	6875	6884	6893	1	2	3	4	4	5	6	7	8
49	6902	6911	6920	6928	6937	6946	6955	6964	6972	6981	1	2	3	4	4	5	6	7	8
50	6990	6998	7007	7016	7024	7033	7042	7050	7059	7067	1	2	3	3	4	5	6	7	8
51	7076	7084	7093	7101	7110	7118	7126	7135	7143	7152	1	2	3	3	4	5	6	7	8
52	7160	7168	7177	7185	7193	7202	7210	7218	7226	7235	1	2	2	3	4	5	6	7	7
53	7243	7251	7259	7267	7275	7284	7292	7300	7308	7316	1	2	2	3	4	5	6	6	7
54	7324	7332	7340	7348	7356	7364	7372	7380	7388	7396	1	2	2	3	4	5	6	6	7

LOGARITHMS.

Natural Numbers.	0	1	2	3	4	5	6	7	8	9	Proportional Parts.								
											1	2	3	4	5	6	7	8	6
55	7404	7412	7419	7427	7435	7443	7451	7459	7466	7474	1	2	2	3	4	5	5	6	7
56	7482	7490	7497	7505	7513	7520	7528	7536	7543	7551	1	2	2	3	4	5	5	6	7
57	7559	7566	7574	7582	7589	7597	7604	7612	7619	7627	1	2	2	3	4	5	5	6	7
58	7634	7642	7649	7657	7664	7672	7679	7686	7694	7701	1	1	2	3	4	4	5	6	7
59	7709	7716	7723	7731	7738	7745	7752	7760	7767	7774	1	1	2	3	4	4	5	6	7
60	7782	7789	7796	7803	7810	7818	7825	7832	7839	7846	1	1	2	3	4	4	5	6	6
61	7853	7860	7868	7875	7882	7889	7896	7903	7910	7917	1	1	2	3	4	4	5	6	6
62	7924	7931	7938	7945	7952	7959	7966	7973	7980	7987	1	1	2	3	3	4	5	6	6
63	7993	8000	8007	8014	8021	8028	8035	8041	8048	8055	1	1	2	3	3	4	5	5	6
64	8062	8069	8075	8082	8089	8096	8102	8109	8116	8122	1	1	2	3	3	4	5	5	6
65	8129	8136	8142	8149	8156	8162	8169	8176	8182	8189	1	1	2	3	3	4	5	5	6
66	8195	8202	8209	8215	8222	8228	8235	8241	8248	8254	1	1	2	3	3	4	5	5	6
67	8261	8267	8274	8280	8287	8293	8299	8306	8312	8319	1	1	2	3	3	4	5	5	6
68	8325	8331	8338	8344	8351	8357	8363	8370	8376	8382	1	1	2	3	3	4	4	5	6
69	8388	8395	8401	8407	8414	8420	8426	8432	8439	8445	1	1	2	2	3	4	4	5	6
70	8451	8457	8463	8470	8476	8482	8488	8494	8500	8506	1	1	2	2	3	4	4	5	6
71	8513	8519	8525	8531	8537	8543	8549	8555	8561	8567	1	1	2	2	3	4	4	5	5
72	8573	8579	8585	8591	8597	8603	8609	8615	8621	8627	1	1	2	2	3	4	4	5	5
73	8633	8639	8645	8651	8657	8663	8669	8675	8681	8686	1	1	2	2	3	4	4	5	5
74	8692	8698	8704	8710	8716	8722	8727	8733	8739	8745	1	1	2	2	3	4	4	5	5
75	8751	8756	8762	8768	8774	8779	8785	8791	8797	8802	1	1	2	2	3	3	4	5	5
76	8808	8814	8820	8825	8831	8837	8842	8848	8854	8859	1	1	2	2	3	3	4	5	5
77	8865	8871	8876	8882	8887	8893	8899	8904	8910	8915	1	1	2	2	3	3	4	4	5
78	8921	8927	8932	8938	8943	8949	8954	8960	8965	8971	1	1	2	2	3	3	4	4	5
79	8976	8982	8987	8993	8998	9004	9009	9015	9020	9025	1	1	2	2	3	3	4	4	5
80	9031	9036	9042	9047	9053	9058	9063	9069	9074	9079	1	1	2	2	3	3	4	4	5
81	9085	9090	9096	9101	9106	9112	9117	9122	9128	9133	1	1	2	2	3	3	4	4	5
82	9138	9143	9149	9154	9159	9165	9170	9175	9180	9186	1	1	2	2	3	3	4	4	5
83	9191	9196	9201	9206	9212	9217	9222	9227	9232	9238	1	1	2	2	3	3	4	4	5
84	9243	9248	9253	9258	9263	9269	9274	9279	9284	9289	1	1	2	2	3	3	4	4	5
85	9294	9299	9304	9309	9315	9320	9325	9330	9335	9340	1	1	2	2	3	3	4	4	5
86	9345	9350	9355	9360	9365	9370	9375	9380	9385	9390	1	1	2	2	3	3	4	4	5
87	9395	9400	9405	9410	9415	9420	9425	9430	9435	9440	0	1	1	2	2	3	3	4	4
88	9445	9450	9455	9460	9465	9469	9474	9479	9484	9489	0	1	1	2	2	3	3	4	4
89	9494	9499	9504	9509	9513	9518	9523	9528	9533	9538	0	1	1	2	2	3	3	4	4
90	9542	9547	9552	9557	9562	9566	9571	9576	9581	9586	0	1	1	2	2	3	3	4	4
91	9590	9595	9600	9605	9609	9614	9619	9624	9628	9633	0	1	1	2	2	3	3	4	4
92	9638	9643	9647	9652	9657	9661	9666	9671	9675	9680	0	1	1	2	2	3	3	4	4
93	9685	9689	9694	9699	9703	9708	9713	9717	9722	9727	0	1	1	2	2	3	3	4	4
94	9731	9736	9741	9745	9750	9754	9759	9763	9768	9773	0	1	1	2	2	3	3	4	4
95	9777	9782	9786	9791	9795	9800	9805	9809	9814	9818	0	1	1	2	2	3	3	4	4
96	9823	9827	9832	9836	9841	9845	9850	9854	9859	9863	0	1	1	2	2	3	3	4	4
97	9868	9872	9877	9881	9886	9890	9894	9899	9903	9908	0	1	1	2	2	3	3	4	4
98	9912	9917	9921	9926	9930	9934	9939	9943	9948	9952	0	1	1	2	2	3	3	4	4
99	9956	9961	9965	9969	9974	9978	9983	9987	9991	9996	0	1	1	2	2	3	3	3	4

ANTILOGARITHMS.

Logarithms.	0	1	2	3	4	5	6	7	8	9	Proportional Parts.								
											1	2	3	4	5	6	7	8	9
.00	1000	1002	1005	1007	1009	1012	1014	1016	1019	1021	0	0	1	1	1	1	2	2	2
.01	1023	1026	1028	1030	1033	1035	1038	1040	1042	1045	0	0	1	1	1	1	2	2	2
.02	1047	1050	1052	1054	1057	1059	1062	1064	1067	1069	0	0	1	1	1	1	2	2	2
.03	1072	1074	1076	1079	1081	1084	1086	1089	1091	1094	0	0	1	1	1	1	2	2	2
.04	1096	1099	1102	1104	1107	1109	1112	1114	1117	1119	0	1	1	1	1	2	2	2	2
.05	1122	1125	1127	1130	1132	1135	1138	1140	1143	1146	0	1	1	1	1	2	2	2	2
.06	1148	1151	1153	1156	1159	1161	1164	1167	1169	1172	0	1	1	1	1	2	2	2	2
.07	1175	1178	1180	1183	1186	1189	1191	1194	1197	1199	0	1	1	1	1	2	2	2	2
.08	1202	1205	1208	1211	1213	1216	1219	1222	1225	1227	0	1	1	1	1	2	2	2	3
.09	1230	1233	1236	1239	1242	1245	1247	1250	1253	1256	0	1	1	1	1	2	2	2	3
.10	1259	1262	1265	1268	1271	1274	1276	1279	1282	1285	0	1	1	1	1	2	2	2	3
.11	1288	1291	1294	1297	1300	1303	1306	1309	1312	1315	0	1	1	1	2	2	2	2	3
.12	1318	1321	1324	1327	1330	1334	1337	1340	1343	1346	0	1	1	1	2	2	2	2	3
.13	1349	1352	1355	1358	1361	1365	1368	1371	1374	1377	0	1	1	1	2	2	2	3	3
.14	1380	1384	1387	1390	1393	1396	1400	1403	1406	1409	0	1	1	1	2	2	2	3	3
.15	1413	1416	1419	1422	1426	1429	1432	1435	1439	1442	0	1	1	1	2	2	2	3	3
.16	1445	1449	1452	1455	1459	1462	1466	1469	1472	1476	0	1	1	1	2	2	2	3	3
.17	1479	1483	1486	1489	1493	1496	1500	1503	1507	1510	0	1	1	1	2	2	2	3	3
.18	1514	1517	1521	1524	1528	1531	1535	1538	1542	1545	0	1	1	1	2	2	2	3	3
.19	1549	1552	1556	1560	1563	1567	1570	1574	1578	1581	0	1	1	1	2	2	3	3	3
.20	1585	1589	1592	1596	1600	1603	1607	1611	1614	1618	0	1	1	1	2	2	3	3	3
.21	1622	1626	1629	1633	1637	1641	1644	1648	1652	1656	0	1	1	2	2	2	3	3	3
.22	1660	1663	1667	1671	1675	1679	1683	1687	1690	1694	0	1	1	2	2	2	3	3	3
.23	1698	1702	1706	1710	1714	1718	1722	1726	1730	1734	0	1	1	2	2	2	3	3	4
.24	1738	1742	1746	1750	1754	1758	1762	1766	1770	1774	0	1	1	2	2	2	3	3	4
.25	1778	1782	1786	1791	1795	1799	1803	1807	1811	1816	0	1	1	2	2	2	3	3	4
.26	1820	1824	1828	1832	1837	1841	1845	1849	1854	1858	0	1	1	2	2	3	3	3	4
.27	1862	1866	1871	1875	1879	1884	1888	1892	1897	1901	0	1	1	2	2	3	3	3	4
.28	1905	1910	1914	1919	1923	1928	1932	1936	1941	1945	0	1	1	2	2	3	3	4	4
.29	1950	1954	1959	1963	1968	1972	1977	1982	1986	1991	0	1	1	2	2	3	3	4	4
.30	1995	2000	2004	2009	2014	2018	2023	2028	2032	2037	0	1	1	2	2	3	3	4	4
.31	2042	2046	2051	2056	2061	2065	2070	2075	2080	2084	0	1	1	2	2	3	3	4	4
.32	2089	2094	2099	2104	2109	2113	2118	2123	2128	2133	0	1	1	2	2	3	3	4	4
.33	2138	2143	2148	2153	2158	2163	2168	2173	2178	2183	0	1	1	2	2	3	3	4	4
.34	2188	2193	2198	2203	2208	2213	2218	2223	2228	2234	1	1	2	2	3	3	4	4	5
.35	2239	2244	2249	2254	2259	2265	2270	2275	2280	2286	1	1	2	2	3	3	4	4	5
.36	2291	2296	2301	2307	2312	2317	2323	2328	2333	2339	1	1	2	2	3	3	4	4	5
.37	2344	2350	2355	2360	2366	2371	2377	2382	2388	2393	1	1	2	2	3	3	4	4	5
.38	2399	2404	2410	2415	2421	2427	2432	2438	2443	2449	1	1	2	2	3	3	4	4	5
.39	2455	2460	2466	2472	2477	2483	2489	2495	2500	2506	1	1	2	2	3	3	4	5	5
.40	2512	2518	2523	2529	2535	2541	2547	2553	2559	2564	1	1	2	2	3	4	4	5	5
.41	2570	2576	2582	2588	2594	2600	2606	2612	2618	2624	1	1	2	2	3	4	4	5	5
.42	2630	2636	2642	2649	2655	2661	2667	2673	2679	2685	1	1	2	2	3	4	4	5	6
.43	2692	2698	2704	2710	2716	2723	2729	2735	2742	2748	1	1	2	3	3	4	4	5	6
.44	2754	2761	2767	2773	2780	2786	2793	2799	2805	2812	1	1	2	3	3	4	4	5	6
.45	2818	2825	2831	2838	2844	2851	2858	2864	2871	2877	1	1	2	3	3	4	5	5	6
.46	2884	2891	2897	2904	2911	2917	2924	2931	2938	2944	1	1	2	3	3	4	5	5	6
.47	2951	2958	2965	2972	2979	2985	2992	2999	3006	3013	1	1	2	3	3	4	5	5	6
.48	3020	3027	3034	3041	3048	3055	3062	3069	3076	3083	1	1	2	3	4	4	5	6	6
.49	3090	3097	3105	3112	3119	3126	3133	3141	3148	3155	1	1	2	3	4	4	5	6	6

ANTILOGARITHMS.

Logarithms.	0	1	2	3	4	5	6	7	8	9	Proportional Parts. 1	2	3	4	5	6	7	8	9
.50	3162	3170	3177	3184	3192	3199	3206	3214	3221	3228	1	1	2	3	4	4	5	6	7
.51	3236	3243	3251	3258	3266	3273	3281	3289	3296	3304	1	2	2	3	4	5	5	6	7
.52	3311	3319	3327	3334	3342	3350	3357	3365	3373	3381	1	2	2	3	4	5	5	6	7
.53	3388	3396	3404	3412	3420	3428	3436	3443	3451	3459	1	2	2	3	4	5	6	6	7
.54	3467	3475	3483	3491	3499	3508	3516	3524	3532	3540	1	2	2	3	4	5	6	6	7
.55	3548	3556	3565	3573	3581	3589	3597	3606	3614	3622	1	2	2	3	4	5	6	7	7
.56	3631	3639	3648	3556	3664	3673	3681	3690	3698	3707	1	2	3	3	4	5	6	7	8
.57	3715	3724	3733	3741	3750	3758	3767	3776	3784	3793	1	2	3	3	4	5	6	7	8
.58	3802	3811	3819	3828	3837	3846	3855	3864	3873	3882	1	2	3	4	4	5	6	7	8
.59	3890	3899	3908	3917	3926	3936	3945	3954	3963	3972	1	2	3	4	5	5	6	7	8
.60	3981	3990	3999	4009	4018	4027	4036	4046	4055	4064	1	2	3	4	5	6	6	7	8
.61	4074	4083	4093	4102	4111	4121	4130	4140	4150	4159	1	2	3	4	5	6	7	8	9
.62	4169	4178	4188	4198	4207	4217	4227	4236	4246	4256	1	2	3	4	5	6	7	8	9
.63	4266	4276	4285	4295	4305	4315	4325	4335	4345	4355	1	2	3	4	5	6	7	8	9
.64	4365	4375	4385	4395	4406	4416	4426	4436	4446	4457	1	2	3	4	5	6	7	8	9
.65	4467	4477	4487	4498	4508	4519	4529	4539	4550	4560	1	2	3	4	5	6	7	8	9
.66	4571	4581	4592	4603	4613	4624	4634	4645	4656	4667	1	2	3	4	5	6	7	9	10
.67	4677	4688	4699	4710	4721	4732	4742	4753	4764	4775	1	2	3	4	5	7	8	9	10
.68	4786	4797	4808	4819	4831	4842	4853	4864	4875	4887	1	2	3	4	6	7	8	9	10
.69	4898	4909	4920	4932	4943	4955	4966	4977	4989	5000	1	2	3	5	6	7	8	9	10
.70	5012	5023	5035	5047	5058	5070	5082	5093	5105	5117	1	2	4	5	6	7	8	9	11
.71	5129	5140	5152	5164	5176	5188	5200	5212	5224	5236	1	2	4	5	6	7	8	10	11
.72	5248	5260	5272	5284	5297	5309	5321	5333	5346	5358	1	2	4	5	6	7	9	10	11
.73	5370	5383	5395	5408	5420	5433	5445	5458	5470	5483	1	3	4	5	6	8	9	10	11
.74	5495	5508	5521	5534	5546	5559	5572	5585	5598	5610	1	3	4	5	6	8	9	10	12
.75	5623	5636	5649	5662	5675	5689	5702	5715	5728	5741	1	3	4	5	7	8	9	10	12
.76	5754	5768	5781	5794	5808	5821	5834	5848	5861	5875	1	3	4	5	7	8	9	11	12
.77	5888	5902	5916	5929	5943	5957	5970	5984	5998	6012	1	3	4	5	7	8	10	11	12
.78	6026	6039	6053	6067	6081	6095	6109	6124	6138	6152	1	3	4	6	7	8	10	11	13
.79	6166	6180	6194	6209	6223	6237	6252	6266	6281	6295	1	3	4	6	7	9	10	11	13
.80	6310	6324	6339	6353	6368	6383	6397	6412	6427	6442	1	3	4	6	7	9	10	12	13
.81	6457	6471	6486	6501	6516	6531	6546	6561	6577	6592	2	3	5	6	8	9	11	12	14
.82	6607	6622	6637	6653	6668	6683	6699	6714	6730	6745	2	3	5	6	8	9	11	12	14
.83	6761	6776	6792	6808	6823	6839	6855	6871	6887	6902	2	3	5	6	8	9	11	13	14
.84	6918	6934	6950	6966	6982	6998	7015	7031	7047	7063	2	3	5	6	8	10	11	13	15
.85	7079	7096	7112	7129	7145	7161	7178	7194	7211	7228	2	3	5	7	8	10	12	13	15
.86	7244	7261	7278	7295	7311	7328	7345	7362	7379	7396	2	3	5	7	8	10	12	13	15
.87	7413	7430	7447	7464	7482	7499	7516	7534	7551	7568	2	3	5	7	9	10	12	14	16
.88	7586	7603	7621	7638	7656	7674	7691	7709	7727	7745	2	4	5	7	9	11	12	14	16
.89	7762	7780	7798	7816	7834	7852	7870	7889	7907	7925	2	4	5	7	9	11	13	14	16
.90	7943	7962	7980	7998	8017	8035	8054	8072	8091	8110	2	4	6	7	9	11	13	15	17
.91	8128	8147	8166	8185	8204	8222	8241	8260	8279	8299	2	4	6	8	9	11	13	15	17
.92	8318	8337	8356	8375	8395	8414	8433	8453	8472	8492	2	4	6	8	10	12	14	15	17
.93	8511	8531	8551	8570	8590	8610	8630	8650	8670	8690	2	4	6	8	10	12	14	16	18
.94	8710	8730	8750	8770	8790	8810	8831	8851	8872	8892	2	4	6	8	10	12	14	16	18
.95	8913	8933	8954	8974	8995	9016	9036	9057	9078	9099	2	4	6	8	10	12	15	17	19
.96	9120	9141	9162	9183	9204	9226	9247	9268	9290	9311	2	4	6	8	11	13	15	17	19
.97	9333	9354	9376	9397	9419	9441	9462	9484	9506	9528	2	4	7	9	11	13	15	17	20
.98	9550	9572	9594	9616	9638	9661	9683	9705	9727	9750	2	4	7	9	11	13	16	18	20
.99	9772	9795	9817	9840	9863	9886	9908	9931	9944	9977	2	5	7	9	11	14	16	18	20

APPENDIX.

ARSENIC.

Determination of Arsenic in Crude Sulphur.—Weigh 10 grams (or more, if considered necessary) of the dry crude sulphur into a tall narrow beaker of about 200 cc. capacity. Add sufficient carbon disulphide to dissolve practically all the sulphur, perhaps 30 to 50 cc. A small amount of the sulphur may exist as a modification insoluble in carbon disulphide. The failure of this to dissolve is usually immaterial. Add 50 cc. of water containing about 7 cc. of strong ammonia. Place the beaker in hot water, repeatedly renewed, until all the carbon disulphide has boiled away. Most of the sulphur precipitates. A little remains in solution as ammonium sulphide, and this solution contains all the arsenic. Filter, washing with warm water. Receive the filtrate in an 8-oz. flask. Add 1 gram of sodium carbonate and boil until all the ammonia is expelled. Now add 3 grams of anhydrous sodium sulphate and 5 cc. of strong sulphuric acid. Boil until all the free acid is expelled and the contents of the flask reduced to a melt. Cool with the flask inclined to prevent the cake from cracking the bottom. Add 50 cc. of hot water and boil to dissolve the cake. Finish by the iodine method, as described in vi, 1. Use an iodine solution of about one-fifth the usual strength. Report in parts per million. If As_2O_3 is required, $As \times 1.32 = As_2O_3$.

Run a blank test by placing in a flask 1 gram of sodium carbonate, 3 grams of sodium sulphate and 5 cc. of sulphuric acid, running the mixture down and finishing as above. Deduct the iodine required for the blank from that used in the analysis before making the calculation for arsenic.

Note.—I have found results obtained by the above method (which takes only one or two hours) to closely check those obtained by much more tedious and elaborate schemes.

COPPER.

Standardizing the Thiosulphate Solution.—In the method of standardizing described on p. 84, pure copper foil is used. This is usually difficult to obtain, necessitating a determination of the actual copper value of the foil used. An equally accurate, and much shorter, method of standardizing is as follows: Place about 100 cc. of water in an 8-oz. flask. Add 5 cc. of glacial acetic acid, 6 cc. of 50 per cent potassium iodide solution, and then run in from a burette about 35 cc. of the usual permanganate solution used for iron (p. 120). Titrate with the thiosulphate solution while the permanganate burette is draining. When near the end add starch solution, and, finally, before finishing, return the flask to the permanganate burette and bring the reading to some definite point. Now complete the titration with the thiosulphate. Multiply the Fe value of the permanganate by 1.139 to obtain the copper value. Multiply this by the number of cubic centimeters of permanganate used, and divide the result by the number of cubic centimeters of thiosulphate used. This gives the standard. All the figures are usually based on 0.5 gram of material taken for assay.

Improved Determination of End-Point.—With small amounts of copper the end-point in the thiosulphate titration is very sharp. With high percentages the exact point requires much care to determine accurately. This is due to the faint purplish tinge of the precipitated cuprous iodide. To neutralize this color, add, at any time during the titration, a little dilute silver nitrate solution, say 2 cc. of a solution of 1 gram of silver nitrate in 200 cc. of water. The precipitated silver iodide, or bromide, is slightly yellowish and tends to neutralize the purplish tinge, thus making the final end-point much sharper.

Improving the Stability of the Thiosulphate Solution.—Solutions made with the sodium thiosulphate I formerly obtained were found to hold their strength very well if kept in an amber glass bottle. More recently, this precaution did not avail. The solution constantly lost strength and daily standardization became necessary. I subsequently discovered that the decomposition was entirely prevented by the addition of a little free sodium or potassium hydroxide, thus indicating that the trouble was due to CO_2. The amber bottle is still used to prevent action of light. In making up the thiosulphate solution I add about 5 grams of sodium hydroxide per liter. This does not interfere with either the copper or lead titration. The solution, instead of losing strength, usually gains slightly in time, probably on account of evaporation.

Sulphur Dioxide Method for Determining Copper Minerals in Partly Oxidized Ores.*—A survey of the status of present laboratory practice brought out the need of a correct and rapid method for the selective determination of the quantity of copper in the sulphide form, on the one hand, and of that in the form of combined oxides, carbonates, silicates, and native or metallic copper, on the other hand, in partly oxidized ores and in mill products from these ores. Such a method is necessary for control determination on flotation mills treating sulphide ores, and also on lixiviation works using either acids or alkalines as the

* Barneveld and Leaver. U. S Bureau of Mines, Technical paper, 198.

active solvent for the copper in oxidized and in silicate form. For convenience, such copper is termed "oxidized copper."

Procedure.—Place 2 grams of pulp, ground to a fineness of 100 to 150 mesh, in a bottle, and 100 cc. of a 3 per cent solution of sulphur dioxide. Seal the bottle and agitate by rolling one-half to two hours. Filter; wash the residue with sulphur dioxide solution; add the washings to the filtrate, which will contain in solution all oxides, carbonates and silicates of copper, and all metallic copper. Add 5 to 10 cc. of nitric acid and boil down to 20 cc. Dilute with distilled water to 150 cc. and determine the copper by the electrolytic method in the usual way.

The residue from the filtration contains the unaltered and undissolved copper sulphides. In the experimental work the copper present as sulphide was separately determined, in order to check the determination of oxidized copper. Ordinarily this step would not be necessary. In analyses of the low-grade porphyry copper ores of the Southwest, the sulphides may be readily decomposed and all the copper dissolved by proceeding as follows: To the residue, add 5 cc. of sulphuric acid and 10 cc. of nitric acid and boil until dense white fumes appear. Add 5 cc. of nitric acid and dilute with distilled water to 150 cc. Determine the copper by the electrolytic method. This method of determining copper in the residue is not suited for heavy sulphide ores containing interfering bases, and for such ores standard methods should be used.

Preparation of Solution.—Although sulphur dioxide solution (sulphurous acid) may be readily purchased, it is decidedly unstable; hence the solution should be prepared in the laboratory as needed. Small quantities are easily made by adding moderately strong sulphuric acid to scrap copper tinned on one side; the resulting sulphur dioxide gas is absorbed in water. For continuous work it is better to purchase liquid sulphur dioxide in steel cylinders and drums, which are obtainable in sizes ranging from 6-lb. to 200-lb. capacity. In the first experiments at the Tucson station the sulphur dioxide gas was introduced directly

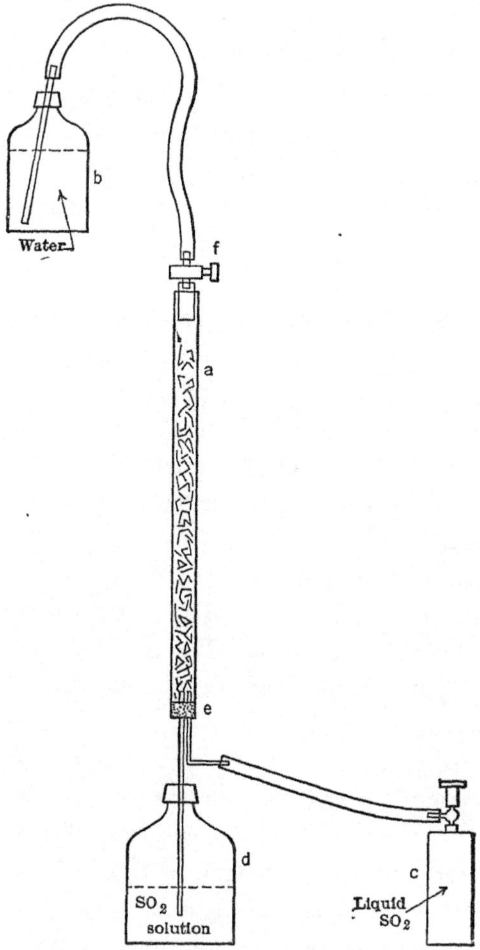

FIG. 1.—Apparatus for Preparation of Sulphur Dioxide Solution: *a*, absorption tower; *b*, distilled water bottle; *c*, container for liquid sulphur dioxide; *d*, bottle for sulphur dioxide solution; *e*, plug of sealing wax.

into the bottle containing the water and pulp. Much loss of gas resulted and the procedure was otherwise unsatisfactory.

Later the simple apparatus shown in Fig. 1 was evolved. A vertical absorption tower *a*, 42 inches long, made of $\frac{3}{4}$-inch to 1-inch glass tubing and filled with broken hard-burned fire clay, is set between two glass bottles *b* and *d*, of 3 to 5 gallons' capacity, the bottle *b* being placed about 5 feet above the other bottle. This tower is open at the top and sealed at the bottom with a plug of sealing wax *e*, through which two small glass tubes extend. The upper bottle *b* contains distilled water, which is siphoned into the upper end of the absorption tower, the flow being regulated by a stopcock *f*. A cylinder *c* (capacity 6 to 50 pounds), containing liquid sulphur dioxide, is connected to one of the glass tubes extending into the absorption tower. On opening the valve of this cylinder the liquid sulphur dioxide issuing from the valve is gasified by the reduction in pressure and passes into the tower, where it is absorbed by the water from bottle *b*, converted into sulphur dioxide solution of the desired strength and caught in the stock bottle *d*.

This apparatus gives entire satisfaction. With little attention, a 3 per cent solution of sulphur dioxide may be produced at the rate of 3 liters per hour. The cylinder containing liquid sulphur dioxide indicated in the sketch may be replaced with an SO_2 gas generator.

Strength of Solution and Time of Contact.—Considerable variation as regards strength of solution and time of contact will be necessary in treating different ores from different localities. In general, for porphyry copper ores a solution containing 3 per cent sulphur dioxide should be used. With some ores much weaker solutions, containing as low as 0.75 per cent sulphur dioxide, will do the work. Merely introducing the pulp into the solution, shaking the bottle for a few minutes, and letting it stand, will not dissolve the copper; constant agitation is essential. For a small number of tests a bottle-agitating machine will give satisfactory results. For analytical work, where large numbers of samples are run, as in a mine laboratory, a bottle-rolling machine will be found more satisfactory, not only for this purpose, but for all solutions requiring constant agitation. The time of contact necessary to completely dissolve the oxidized copper minerals was found to vary from one-half to two hours. Most of the ores and products tested gave complete recovery in half an hour, and the most refractory ores yielded in less than two hours.

Testing Strength of Solution.—To determine the strength of the sulphur dioxide solution, the following adaptation of a well-known reaction is recommended. It is based on the fact that introducing either weak or concentrated sulphurous acid into a solution of iodine will result in the complete oxidation of the sulphur dioxide.

Prepare an iodine solution by dissolving 16.8 grams of potassium iodide in distilled water, adding 8.4 grams of pure resublimed iodine, and shaking until the iodine is completely dissolved. The more concentrated the potassium iodide solution, the more readily will the iodine dissolve. Bring the solution to proper strength by adding enough distilled water to make a volume of 1 liter. Then standardize the solution by the thiosulphate method, using starch indicator.

The determination is made as follows: To a measured quantity of standard iodine solution add slowly, with constant stirring, the proper volume of sulphur dioxide solution. So regulate the volume of iodine solution used that the mixture always contains a decided excess of iodine over the quantity required to oxidize the sulphur dioxide being added. An excess of sulphur dioxide causes the solution to clear and to lose its dark red color. If an excess of sulphur dioxide is added, the determination is spoiled, and the test should be repeated with fresh sulphur dioxide solution and a larger quantity of iodine solution. Thus, there is a relation between the strength and quantity of sulphur dioxide solution and the quantity of standard iodine solution. In general, when the solution to be standardized varies in strength from 1 to 3 per cent sulphur dioxide, add 1 cc. of this solution to 20 cc. of standard iodine solution. When the sulphur dioxide solution is appreciable below 1 per cent in strength add a larger quantity of it. The mixture is then titrated by the thiosulphate method to determine the quantity of iodine remaining in the mixture. The difference between this quantity and the total quantity of iodine represents the iodine used in oxidizing the sulphur dioxide. The strength of the sulphur dioxide solution may then be calculated according to the formula:

$$SO_2+H_2O+2I=2HI+SO_3.$$

Metallic iron, from the grinding, etc., in amount up to 2 per cent is without influence on the results. The amount found in finely ground ores is usually less than 1 per cent. An essential requirement is the continued presence of a strong excess of SO_2, which readily dissolved the metallic iron. In order to insure this excess, the SO_2 must be introduced in the form of sulphur dioxide solution. If SO_2 is introduced in the form of a gas, unstable conditions arise from the unequal distribution of the gas; then the presence of metallic iron, in quantities less than 2 per cent, may cause the precipitation of copper in the form of cement copper, cupro-cupric sulphites, and probably also as complex sulphides, which are not redissolved.

FLUOR SPAR.

4. **Rapid Practical Method for Fluor Spar.**—Of the methods in the text I have found Kneeland's the more satisfactory. Penfield's method promises well, but appears to be very uncertain. A practical method, giving fairly accurate results in most cases, is as follows:

Take 0.5 gram of the ore in an 8-oz. flask, moisten with water, add 5 cc. of glacial acetic acid and boil nearly to pastiness. Take up in about 30 cc. of equal parts of glacial acetic acid and water and boil gently for a few minutes. These operations will not affect calcium fluoride, but will usually extract practically all other calcium salts. Filter, washing well with hot water. Place about 3 grams of powdered anhydrous sodium sulphate in a small platinum dish, mixing in also a little potassium nitrate, if reducible metals are liable to be present. Lay the filter and residue upon this mixture and ignite gently until the paper is burned fairly well. Now cool and add 5–6 cc. of strong sulphuric acid. Heat carefully, to avoid spattering, first to strong fumes, and then to a melt, if possible. If the mass solidifies at the end, without melting, cool sufficiently, add a little more sulphuric acid and heat again. This will usually effect complete decomposition of the fluoride, and expulsion of the fluorine, even if the mass is not completely melted. Allow to cool, cover the dish and dissolve the cake by warming with sufficient water acidulated with 5 cc. of hydrochloric acid. Transfer the solution to the original flask, first filtering, if there is an appreciable amount of insoluble residue. Dilute to about 150 cc. with hot water and proceed with the determination of CaO as described in X, 1, at the same point. Multiply the percentage of CaO found by 1.392 to obtain the percentage of CaF_2.

MANGANESE.

(Referred to from p. 167.)

10a. Determination of MnO_2 Only.—Place 0.5 gram of the finely ground ore in an 8-oz. flask. Add 25 cc. of water, 5 cc. of strong sulphuric acid and as much of the usual standard oxalic acid solution, delivered from a burette, as may be judged necessary to decompose the MnO_2 present and leave a moderate excess of oxalic acid. More may be added later if found necessary. Boil gently until the decomposition of all MnO_2 is effected, adding more water or more oxalic acid solution as required. Finally, dilute, if necessary, with hot water, until the flask is about half full, and titrate with permanganate as usual. Make the same calculation as in **6.** Multiply the per cent of manganese shown by 1.581 to obtain the per cent of MnO_2.

MOLYBDENUM.

1. High-Grade Ores and Concentrates.—Take 0.5 gram of the finely ground material. Place in a thin spun-iron crucible of about 25–30 cc. capacity and mix with 1 gram of sodium carbonate. Now add about 6 grams of sodium peroxide and again mix well. Fuse over a Bunsen burner, using a low flame at first, and gradually bring to dull redness. Rotate the crucible with the tongs, as the mixture melts, so as to obtain a perfect fusion without overheating and destroying the crucible. Allow to cool to a crust on top and then set the crucible in a 400 cc. beaker containing about $\frac{3}{4}$ of an inch of water. Cover the beaker and then upset the crucible with a glass rod. Disintegration usually occurs quickly without further heating. Remove and rinse the cover and then lift out the crucible with the glass rod and wash it. Add 10 grams of ammonium carbonate, which should be sufficient to neutralize all the fixed alkali and leave an excess. Warm until the ammonium carbonate is all dissolved and then filter (best with a Witts plate and suction) and wash at least ten times with hot water. Add about 2 grams of tartaric acid or a tartrate (Rochelle Salts) to the filtrate, which should still remain strongly alkaline. This is to prevent precipitation of tungsten or vanadium in the subsequent acidification. Saturate the alkaline liquid with hydrogen sulphide, and then make slightly acid with diluted (1 : 1) sulphuric acid. Filter off the molybdenum sulphide and wash 5 or 6 times with hot water. The filtrate may still contain a little molybdenum. Make it strongly alkaline with ammonia and again saturate with hydrogen sulphide, then acidify and filter as before, using a second filter. Place both filters and their contents in a large beaker and add 20 cc. of water, 10 cc. of strong nitric acid and about 2 grams of potassium chlorate. Cover the beaker and warm until solution of the molybdenum is complete. Rotate the liquid occasionally, to effect complete solution, but avoid vigorous stirring or such prolonged heating or concentration as would reduce the filters to a fine pulp, which might be very difficult to filter. Dilute somewhat and then filter into an 8-oz. flask, washing the paper residue ten times with hot water. Add 5 cc. of strong sulphuric acid to the filtrate and boil down to strong fumes. This should oxidize all organic matter dissolved from the filters and expel the nitric acid. After cooling, take up the residue in about 150 cc. of warm water and add 3 cc. of a 4 per cent solution of copper sulphate. Now cautiously add, a little at a time at first, 5 grams of 30-mesh zinc and cover the flask with a watch-glass. If the action is too violent, cool the flask to prevent foaming over. Later, the flask may be warmed if necessary. Allow the action to proceed, warm, for about fifteen minutes. This should precipitate all the copper, and, incidentally, any arsenic present. Zinc alone, without added copper, will not remove all the arsenic. Now add a little dilute sulphuric acid, to make certain that the liquid is still acid, and filter. Wash the residue with cold water. To the filtrate add 20 cc. of 1 : 1 sulphuric acid. Pour the hot solution through the reductor in accordance with the accompanying directions (4), which are to be further followed to the end. The Fe factor of the permanganate multiplied by 0.5725 (log. 9.7578) gives the Mo factor. Mo multiplied by $10/6 = MoS_2$.

2. In the absence of tungsten and vanadium the process may be shortened as follows: Boil the filtrate from the fusion down to about 150 cc. and filter off the (usually slight) precipitate that forms. Receive the filtrate in an 8-oz. flask, make slightly acid with sulphuric acid and then add 5 cc. of the strong acid in excess. Again boil down to about 150 cc., add 3 cc. of the copper sulphate solution and finish as above.

3. **Low-Grade Ores.**—The fusion method of decomposition, described above, is the best for concentrates, which contain little silica. With ores, especially low grade, and tailings, where, perhaps a large amount has to be taken to insure accuracy, the dissolved silica may be very troublesome. It is best, in such cases, to decompose with acids in a flask, taking from 1 to 5 grams of substance, as may be deemed advisable, and acids in proportion. Nitric acid may be used first, then add hydrochloric acid, and perhaps more nitric, and boil until all dark particles have disappeared. Molybdenum concentrates might resist the acid treatment for a long while, but with finely ground ores it usually works well. Finally, boil to very small bulk, dilute with hot water, add an excess of ammonia, boil, filter and continue as above.

4. Reduction of the Molybdenum Solution.*—The reductor tube should have at least an 8-inch zinc column, ⅝-inch in diameter, using 20–30-mesh amalgamated granulated zinc. The end of the reductor tube should be prolonged to reach nearly to the bottom of the receiving flask. The flask (32-oz.) should contain 30–35 cc. of ferric phosphate solution (see below) and the end of the reductor tube should dip below this, so that the reduced molybdenum solution is not exposed to the air, but is oxidized at the expense of the ferric phosphate. The molybdenum solution should contain about 5 cc. of free sulphuric acid per 100 cc. and should be passed rather slowly (about five minutes) through the reductor, at a temperature of 50°–75° C. After passing the molybdenum solution, the reductor should be washed by passing 100 cc. of hot water containing 5 cc. of strong sulphuric acid, followed by hot water alone. The titration with the permanganate (of the usual strength used for iron) should be made in the warm solution immediately after reduction. With sufficient molybdenum present, the liquid in the receiving flask is red. As permanganate is added, the red slowly fades and the solution becomes colorless; then the final pink tint which follows is easily recognized. A blank should be made, beginning with 150 cc. of water and 5 cc. of sulphuric acid in an 8-oz. flask, adding the copper sulphate and zinc as in 1 and continuing as there described. The usual correction obtained in this way is about 0.20 cc. of permanganate solution.

Under the above conditions the reduction is from MoO_3 to Mo_2O_3, and is exact. The permanganate is standardized against sodium oxalate (Bureau of Standards), or oxalic acid (see Chapter on Iron), making it the same strength as for iron determinations.

---

* Reduction in the Permanganate Method for Molybdenum. Gooch, Methods in Chemical Analysis. Wiley and Sons, 1912. Pp. 424–429.

Ferric Phosphate Solution.—Dissolve 100 grams of Ferric Ammonium Sulphate in 500 cc. of water, adding 25 cc. of strong sulphuric acid, followed by 150 cc. of syrupy (85 per cent) phosphoric acid, and diluting to 1000 cc.

PHOSPHORUS.

Method for Phosphorus in Tungsten Ores.—Take 1 gram of the ore and fuse in a platinum dish with 5 grams of sodium carbonate and 1 gram of sodium or potassium nitrate. Disintegrate with hot water and filter, washing with hot water. If much silica is present, acidify the filtrate with hydrochloric acid, evaporate to dryness, take up in hydrochloric acid and water, in the usual way for silica, and filter, washing with dilute hydrochloric acid. If there is but little silica, its removal may be omitted and the filtrate from the fusion simply made acid with hydrochloric acid. To the acid solution (which may contain a precipitate of WO_3) add sufficient ferric chloride solution to color the liquid plainly yellow, avoiding a great excess. Now make strongly alkaline with ammonia, boil and filter, washing well with hot water. Dissolve the precipitate, after rinsing most of it back into the beaker, with hydrochloric acid, pouring the solution through the same filter, and wash out every trace of iron with warm dilute hydrochloric acid. Dilute the filtrate somewhat with hot water and re-precipitate the iron with ammonia as before. Filter, washing the precipitate at least ten times with hot water. Dissolve the precipitate on the filter with a hot mixture of 10 cc. of strong nitric acid and 20 cc. of water, washing out every trace of iron with hot, dilute nitric acid. Receive the filtrate in an 8-oz. flask. Boil to about 40 cc., or less, if the bulk is larger. Precipitate the phosphorus with molybdate solution in the usual way (p. 206). Filter through a weighed Gooch crucible. Mul tiply the weight of the yellow precipitate by 0.0163 to obtain the phosphorus.

POTASH.

1. **Short Perchlorate Method for Soluble Potash.**—In the case of water, weigh 50 grams, more or less, according to supposed richness, and evaporate to dryness in a small weighed porcelain or platinum dish. Heat the residue for some time at about 150° C., cool, weigh. This gives the per cent of salts in the water. If the weight is small, 0.5 gram or less, all may be used and dissolved right in the dish. If considerably more than 0.5 gram grind to powder in the dish with an agate pestle, and weigh out 0.5 gram for the analysis. Continue as described below.

2. In the case of salts, such as mixtures of sulphates, chlorides and carbonates, have the material finely ground and well mixed. Using a weighing-bottle, weigh something more than 0.5 gram into it. Dry this at 150° C. This gives the moisture and the weight of dry substance. Wash the material into a small beaker, add a drop of methyl orange, cover the beaker and make just acid with hydrochloric acid. If filtration appears advisable, boil a moment if CO_2 has been shown and then filter through a small filter into a beaker. Wash six times with hot water. Dilute the filtrate, or boil it down, as may be necessary, to 50 cc., and add an excess of 3/20 cc. of strong hydrochloric acid. Heat to boiling and add 10 per cent barium chloride solution in slight excess. Take time to do this carefully, as a large excess of barium chloride is troublesome in subsequently converting to perchlorate. The hot liquid may be filtered immediately into a shallow evaporating-dish, washing with hot water. Any small amount of barium sulphate that has failed to separate will do no harm. Add 5 cc. of 20 per cent perchloric acid and evaporate to

dryness on a hot-plate and until fumes of perchloric acid have ceased. Cool sufficiently, wash down the sides of the dish with just enough hot water to dissolve the salts, add 1 or 2 cc. more of perchloric acid and again evaporate as before. If no fumes of perchloric acid appeared on the first evaporation, add 5 cc. for the second evaporation and again repeat with 1 or 2 cc. Allow to become completely cold, then add 25 cc. of a saturated solution of potassium perchlorate in 95 per cent grain alcohol (or denatured alcohol). Cover the dish and allow a little time for the soluble perchlorates to dissolve. The final salts consist of the perchlorates of the bases present, of which only potassium perchlorate is insoluble in the alchohol, mixed perhaps with barium sulphate and small amounts of organic and other insoluble matter.

Prepare a Gooch crucible with a thick asbestos mat and have it dry. Thoroughly loosen up the salts in the dish with a rubber-tipped glass rod, rubbing the particles so as to leave only a powdery residue, free from lumps. Rub a little grease on the lower side of the lip of the dish, as the alcohol has a tendency to run down the outside, and filter through the Gooch (not yet weighed). Using the alcoholic perchlorate solution in a small wash-bottle, transfer all the residue from the dish to the filter, allow all the liquid to run through, and then wash six times with the alcoholic perchlorate solution, allowing to drain completely each time. Finally, drain thoroughly and then dry in an oven at 130°–150° C., cool and weigh. Now wash the residue in the Gooch six times with hot water, which will dissolve the potassium perchlorate, then twice with alcohol, and once with a very little of the alcoholic perchlorate solution. Drain completely, dry, cool and weigh as before. The loss in weight represents the potassium perchlorate. Multiply this by 0.3399 to obtain the K_2O.

The alcoholic perchlorate solution is made by simply shaking up an excess of potassium perchlorate with the alcohol, allowing to stand some time, best over night, and always shaking the mixture and filtering or decanting a portion before use, so as to have the solution saturated at the laboratory temperature when used.

SULPHUR.

Weight 0.5 gram of the finely ground substance. Mix in a small spun-iron crucible (about 25–30 cc. capacity) with 1 gram of dry sodium carbonate and about 5 grams of sodium peroxide. Use peroxide of calorimeter quality, and, after trial, it may be measured instead of weighed. Place the crucible in a hole cut in asbestos board, to prevent absorption of sulphur fumes, and fuse at a gentle heat over a Bunsen burner. While still quite hot set the crucible in about half an inch of water contained in a No. 4 beaker, cover the beaker and agitate it so as to overturn the crucible. The contents will quickly disintegrate. Remove and wash the crucible. Add about 5 grams of ammonium carbonate to partially destroy the causticity of the peroxide, warm gently until dissolved, and then filter. I use an 11-cm. filter with a small wad of absorbent cotton in the point, which facilitates the filtration. Wash at least ten times with hot water. Receive the filtrate in a 600-cc. Erlenmeyer flask. If the filtrate is greenish, indicating manganese, it is a good plan to add 5 cc. of alcohol, boil and then filter again. Add a drop or two of methyl orange as indicator, make slightly acid with hydrochloric acid and then about 1.2 cc. of the strong acid in excess. Dilute to 400 cc. with hot water, heat to boiling and add an excess of 10 per cent. barium chloride solution through a small funnel with the stem drawn out to a pin-hole, so as to deliver a very small stream. Continue the boiling for a few minutes and then allow to stand, hot, for several hours. Finally, filter through a double 11-cm. ashless filter, wash ten times with hot water and ignite as usual for $BaSO_4$.

Free Sulphur in Ores, Etc.—Have the material thoroughly air-dried. Heating to expel moisture is liable to volatilize sulphur. Weigh from 1 to 5 grams, according to amount of free sulphur possibly present. Place in a 200 cc. beaker, add about 30 cc. of aniline and heat the mixture just to the boiling-point for a few minutes, frequently rotating the beaker to stir the contents. Pour the hot liquid through a weighed Gooch crucible with asbestos mat, using suction. Rinse out the beaker several times with carbon disulphide, pouring through the residue in the crucible, and until the latter is washed free from aniline. A slight residue remaining in the beaker will soon dry and may be brushed into the crucible. Finally, heat the latter to constant weight at about 100° C. The difference between the final weight of the residue and the original weight of substance taken represents the free sulphur.

Carbon disulphide alone cannot be depended upon to dissolve all the free sulphur, as a modification of the element that is insoluble in carbon disulphide is liable to be present.

It is best to have some water in the receiver under the Gooch filter, both to prevent the hot aniline from cracking the glass, and to prevent the separated sulphur from adhering to the sides.

A possible source of error is water remaining in the substance that might count as free sulphur.

TUNGSTEN.

1. Weigh 0.5 gram of the very finely divided ore into an 8-oz. flask ("copper-flask"). Add 3 grams of anhydrous sodium sulphate and 5 cc. of strong sulphuric acid. Heat over a free flame, best with the flask in a holder, until the free acid is all expelled and the mixture reduced to a red-hot melt. Continue the fusion until the decomposition is complete. By holding the flask at an angle, and giving it a slight circular motion, so that most of the heat is on the curve just above the bottom, the latter is prevented from bulging. Rotate the flask as it cools, so as to distribute the melt on the sides, and thus avoid cracking. The decomposition is usually very quickly effected without any material injury to the flask. When the melt is cool, add 20 cc. of hot water, 20 cc. of strong hydrochloric acid and 5 cc. of strong nitric acid. Boil until the bulk is reduced to about 10 cc. Add 25 cc. of hot water, 3 cc. of cinchonine solution, and cool under the tap to room temperature, or cooler. The cinchonine solution is made by dissolving 25 grams of cinchonine in 200 cc. of 1 : 1 hydrochloric acid. Filter the cold mixture through a 9-cm. filter, returning the first portion if at all cloudy. The flask may be washed out with cold water if done quickly, while the filter still contains liquid, remembering, however, that cold water alone, with no cinchonine present, will soon cause a cloudy filtrate. Wash filter and precipitate twice with dilute cinchonine solution (5 cc. of the above solution diluted to 100 cc.) contained in a small wash-bottle. The flask may retain a little WO_3 adhering to the sides. Spread the filter on a watch-glass and wash the contents, through a funnel, back into the flask with hot water, using as little as possible. Pour a little ammonia upon the filter in the watch-glass and set it on the hot-plate to dry. Add 1 gram of sodium carbonate to the mixture in the flask and boil

to small bulk while the filter is drying. Re-fold the dry filter in the original creases and replace in the funnel. Rub up the adhering residue on the watch-glass with a little water and wash through the filter into a clean flask like the first. Pour the concentrated liquid in the original flask through the filter and wash out the flask with hot water. Now pour a little ammonia into the washed flask, add about 50 cc. of hot water, and use this solution to wash the filter at least ten times, with about 5 cc. each time. Boil the filtrate to small bulk, add 2 grams of anhydrous sodium sulphate and (cautiously) 5 cc. of strong sulphuric acid. The boiling has expelled all the ammonia. Boil the mixture in the flask in the original manner until the free acid is expelled and any carbon from cinchonine is burned off. Cool as before, add 25 cc. of strong hydrochloric acid and 5 cc. of strong nitric acid and boil down to about 10 cc. Add 25 cc. of hot water, 3 cc. of cinchonine solution and cool and filter as before, using an 11-cm. ashless filter. Before washing the filter, pour a little ammonia into the flask to dissolve adhering WO_3 and boil the solution until all free ammonia is expelled. Now add a few drops of hydrochloric acid and about 1 cc. of the dilute cinchonine solution and cool under the tap. Pour the cold mixture into the filter and wash out the flask quickly with cold water. Wash filter and precipitate at least ten times with the cold dilute cinchonine solution. Transfer the washed filter and precipitate to a weighed platinum dish and ignite until the carbon is burned off. Cool, add a few cubic centimeters of hydrofluoric acid and evaporate to dryness to remove any silica. Again ignite and then cool and weigh as WO_3.

Note.—The cinchonine may be replaced by quinine if desired.

Watt's Method.*—Usually 1 gram of the agate-ground sample is used for analysis, whatever the grade, except in the

* Communicated by Hugh F. Watts, Boulder, Colo.

case of extremely low-grade ores or tailings, when 2 grams are taken.

The method of attack by acid depends somewhat on the grade of the ore; if not over 30 per cent it is treated directly by evaporating with about 40 cc. of aqua-regia, if much above this, and always in the case of concentrates, the assay is started with 40 to 50 cc. of hydrochloric acid alone. In the latter case, when the solution has evaporated to say, 15 cc., about 25 cc. of aqua-regia are added and the evaporation again carried on until the solution has evaporated to 15 cc. The evaporation is conducted in a small covered beaker on a hot-plate.

The beaker is now removed from the hot-plate, 50 cc. of hot water are added and the mixture allowed to stand for twenty or thirty minutes. The nearly clear solution is now decanted through a filter, keeping as much of the residue as possible in the beaker. The residue is now washed twice by decantation, using 50 cc. of hot water, containing a little hydrochloric acid each time.

To the residue in the beaker add about 15 cc. of ammonia solution (made by adding 400 cc. of strong ammonia to 2000 cc. of water containing 20 cc. of strong hydrochloric acid). Warm slightly until all the liberated tungstic acid is in solution. Decant through the original filter, receiving the filtrate in a large porcelain crucible. (If the ore is low grade platinum should always be used.) The residue is now examined for any undecomposed particles of mineral; if none are found, transfer the whole to the filter with the ammonia solution (used warm in a wash-bottle). The filter is then further washed with the same solution until free from tungstic acid. Five washings of 2 or 3 cc. each will suffice, if the stream is directed around the top of the filter.

Should any undecomposed mineral be found after the first

treatment with ammonia, as may sometimes be the case owing to the protective action of the liberated tungstic acid, wash three times by decantation with ammonia solution and then, instead of transferring to the filter, again treat with aqua-regia, which should not be stinted in amount, even if the residue is small. Use at least 30 cc. and treat exactly as at first, except that smaller amounts of wash solutions will suffice. The tungstic acid liberated by the second acid treatment is apparently somewhat more difficultly soluble in ammonia than the first obtained, but by warming and stirring it will finally dissolve. Add this solution to the main ammonia solution in the large porcelain crucible.

Evaporate to dryness in the crucible, ignite gently to expel ammonia salts, and finally over the full flame of a good Bunsen or Teclu burner. Cool and weigh as WO_3

If the ore is of low grade, the evaporation should be made in a platinum dish, and the residue, after ignition, moistened with hydrofluoric acid and again ignited before weighing as WO_3.

URANIUM.

I usually employ the following method, which is considerably shorter than that given in the text and apparently quite as accurate.

Take 0.5 gram of the finely ground ore, or more, according to richness. Treat, in an 8-oz. "copper flask," with nitric or hydrochloric acid, or both, to effect complete solution of the uranium. With an ore containing much galena, it is best to start with a mixture of 1 part nitric acid and 2 parts water, adding hydrochloric acid later. Finally, boil to approximate dryness, cool, add 3 cc. of nitric acid, 50 cc. of hot water, and see that everything soluble is dissolved. Now make slightly alkaline with ammonia, then just acid with nitric acid, and again alkaline with a little solid ammonium carbonate. Add about 5 cc. of strong ammonia at this point and 3-4 grams additional of ammonium carbonate. Boil for about a minute, then filter, washing several times with hot water. It is best to use a Witts plate and suction. Boil and concentrate the filtrate in a covered beaker during the next step. Dissolve the precipitate on the filter with a little hot dilute nitric acid, receiving the filtrate in the original flask. Again neutralize and precipitate as before, washing this second precipitate well with hot water. Add the filtrate to the first one and continue the concentration to 150–200 cc. Now acidify with nitric acid, and then, in case of doubt, test for vanadium by adding about 1 cc. of hydrogen peroxide. A reddish-brown color indicates vanadium.

A. Vanadium Present.—Boil to expel any remaining CO_2, make just alkaline with ammonia, then just acid with nitric acid, finally adding about 4 cc. of the latter in excess. A bit of litmus paper in the liquid may be used as indicator, if necessary. Now add 1 gram of lead acetate crystals and then sufficient ammonium acetate solution (about 20 cc.) to neutralize the nitric acid and precipitate the lead vanadate. Boil for about ten minutes and then filter through a close filter, returning the first portion if not perfectly clear. Wash with hot water. Receive the filtrate in a large beaker. If bulky, boil down to perhaps 200–250 cc. Now add ammonia in marked excess and boil for a minute or two to expel any CO_2. Filter hot, paying no attention to a turbid filtrate unless it is yellowish. No washing required. Place the last beaker under the funnel and fill the latter with a strong hot solution of ammonium carbonate, to which some free ammonia has been added. Usually one filling is sufficient to dissolve all the uranium and leave a white lead residue, perhaps slightly discolored by a trace of iron. Wash with hot water, using a little more of the ammonium carbonate solution, if apparently necessary. Add to the filtrate sufficient strong hydrogen sulphide water to precipitate all the remaining lead, or pass the gas for a short time. This also removes traces of iron. Filter, washing with hydrogen sulphide water containing some ammonium carbonate. Boil to expel the sulphide, then acidify with nitric acid and boil off all CO_2. Continue according to C.

B. Vanadium Absent.—Boil the nitric acid solution sufficiently to expel all CO_2, then add ammonia in marked excess and boil a little longer to expel any CO_2 in the ammonia. Filter the hot mixture, returning the first portion if not perfectly clear. No washing required. Dissolve the uranium on the filter with hot ammonium carbonate solution, as described in the last paragraph, and continue from this point as in the same situation above. Do not omit the hydrogen sulphide treatment, for, even in the absence of lead, there will usually be traces of iron to be removed. Continue according to C.

C. Add ammonia in marked excess, boil for about a minute and then filter through an ashless filter, returning the first portion if not clear. Ignite filter and precipitate thoroughly in a porcelain crucible and weigh as U_3O_3. Impurities are usually present. Dissolve the residue in the crucible by warming with a little nitric acid. Dilute and test for vanadium with hydrogen peroxide. A faint brownish tinge may be neglected. Filter off any undissolved residue through a small ashless filter, ignite in the original crucible, weigh and deduct the weight from that of the impure U_3O_3 previously found.

Ammonium Acetate Solution.—Eighty cc. of strong ammonia, 100 cc. of water and 70 cc. of 99 per cent acetic acid.

Note.—A yellow filtrate from the ammonium uranate indicates incomplete precipitation. This may be due to a deficiency in ammonium salts, as ammonium uranate is perceptibly soluble in pure water. Add a gram or so of ammonium nitrate to the filtrate, boil and refilter. Or, better, dissolve the precipitate on the filter with dilute nitric acid, so that the mixed filtrates will be markedly acid, and repeat the precipitation with ammonia. The filtrate should be colorless.

INDEX.

A

D

E

F

G

H

I

J

K

L

M

N

O

P

R

S

V

W

Z

Wiley Special Subject Catalogues

For convenience a list of the Wiley Special Subject Catalogues, envelope size, has been printed. These are arranged in groups—each catalogue having a key symbol. (See special Subject List Below). To obtain any of these catalogues, send a postal using the key symbols of the Catalogues desired.

1—Agriculture. Animal Husbandry. Dairying. Industrial Canning and Preserving.

2—Architecture. Building. Masonry.

3—Business Administration and Management. Law. Industrial Processes: Canning and Preserving; Oil and Gas Production; Paint; Printing; Sugar Manufacture; Textile.

CHEMISTRY

4a General; Analytical, Qualitative and Quantitative; Inorganic; Organic.

4b Electro- and Physical; Food and Water; Industrial; Medical and Pharmaceutical; Sugar.

CIVIL ENGINEERING

5a Unclassified and Structural Engineering.

5b Materials and Mechanics of Construction, including; Cement and Concrete; Excavation and Earthwork; Foundations; Masonry; [illegible]

5c Railroads; Surveying.

5d Dams; Hydraulic Engineering; Pumping and Hydraulics; Irrigation Engineering; River and Harbor Engineering; Water Supply.

(Over)

CIVIL ENGINEERING—*Continued*

5e Highways; Municipal Engineering; Sanitary Engineering; Water Supply. **Forestry. Horticulture, Botany and Landscape Gardening.**

6—Design. Decoration. Drawing: General; Descriptive Geometry; Kinematics; Mechanical.

ELECTRICAL ENGINEERING—PHYSICS

7—General and Unclassified; Batteries; Central Station Practice; Distribution and Transmission; Dynamo-Electro Machinery; Electro-Chemistry and Metallurgy; Measuring Instruments and Miscellaneous Apparatus.

8—Astronomy. Meteorology. Explosives. Marine and Naval Engineering. Military. Miscellaneous Books.

MATHEMATICS

9—General; Algebra; Analytic and Plane Geometry; Calculus; Trigonometry; Vector Analysis.

MECHANICAL ENGINEERING

10a General and Unclassified; Foundry Practice; Shop Practice.

10b Gas Power and Internal Combustion Engines; Heating and Ventilation; Refrigeration.

10c Machine Design and Mechanism; Power Transmission; Steam Power and Power Plants; Thermodynamics and Heat Power.

11—Mechanics.

12—Medicine. Pharmacy. Medical and Pharmaceutical Chemistry. Sanitary Science and Engineering. Bacteriology and Biology.

MINING ENGINEERING

13—General; Assaying; Excavation, Earthwork, Tunneling, Etc.; Explosives; Geology; Metallurgy; Mineralogy; Prospecting; Ventilation.

Printed in Dunstable, United Kingdom

91219913R00231